# Edible Wild Fruits
# and Nuts of Canada

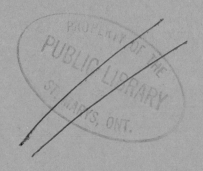

Published by the National Museums
of Canada

Managing editor
Viviane Appleton

Editor
Penny Williams

Production
Barry P. Boucher

Series Design Concept
Eskind Waddell

Layout
Gregory Gregory Limited

Typesetting
Mono Lino Typesetting Company Limited

Printing
Thorn Press Limited

Publications in the Edible Wild Plants of
Canada series:
1  *Edible Garden Weeds of Canada*, 1978
2  *Wild Coffee and Tea Substitutes of Canada*,
   1978
3  *Edible Wild Fruits and Nuts of Canada*,
   1979
4  *Edible Wild Greens of Canada*, 1980

Cette collection existe en français sous
le titre, Plantes sauvages comestibles
du Canada:
1  *Mauvaises herbes comestibles de nos jardins*,
   1978
2  *Succédanés sauvages du thé et du café
   au Canada*, 1978
3  *Fruits et noix sauvages comestibles du
   Canada*, 1979
4  *Verdure sauvage comestible du Canada*,
   1980

Cover: high-bush cranberry (*Viburnum opulus*
var. *americanum*)

Nancy J. Turner
Adam F. Szczawinski

# Edible Wild Fruits and Nuts of Canada

National Museum
of Natural Sciences

National Museums
of Canada

©National Museums of Canada 1979

National Museum of Natural Sciences
National Museums of Canada
Ottawa, Canada  K1A 0M8

Catalogue No. NM95-40/3

Printed in Canada

English edition
ISBN 0-660-00128-4
ISSN 0705-3967

French edition
ISBN 0-660-00129-2
ISSN 0705-3975

# Contents

## Illustration Credits

Oldriska Ceska did the drawings reproduced on pages 23, 31, 40, 48, 53, 64, 73, 76, 77, 80, 101, 105, 127, 142, 160, 165, 181, and 186. Marcel Jomphe did those on pages 115, 119, and 138.

Silvija Ulmanis took the photograph that appears on the cover and on page 60. She also gave generously of her time searching for other colour illustrations to use in this book. The photographers who supplied slides for the colour plates are listed in alphabetical order, with the page numbers on which their photographs occur:

Andrew M. Brown, 95 bottom,
Mary Ferguson, 111 top right,
Sharon Godkin, 45 left,
Eric Haber, 45 right,
Mary I. Moore, 176 bottom left, 190,
J. Renfroe, 152,
Hans Roemer, 111 top left,
Robert D. Turner and Nancy J. Turner, 34, 68, 86, 95 top, 111 bottom, 157, 176 top and bottom right.

Of the photographs obtained from the National Collection of Nature Photographs, National Museums of Canada, J. Lotochinski took those reproduced on pages 91 and 133, and James L. Parker those on pages 170 and 176 top left.

The Department of Fisheries and the Environment, Ottawa, provided the photographs that appear on pages 124 and 147.

# Acknowledgements

We should like to express our sincere thanks to several people whose support and co-operation helped us in the preparation of this book: Dr. Douglas Leechman, now retired in Victoria, formerly of the National Museum of Man, who allowed us to use his extensive notes on edible plants in Canada; Dr. T. Kuiper-Goodman, Toxicological Evaluation Division, Department of Health and Welfare, Ottawa, who so generously reviewed the toxicity of the various species discussed in this book; the Food Advisory Division, Agriculture Canada, which tested some of the recipes included; Dr. W. G. Dore, Ottawa; Mary I. Moore, Petawawa Forest Experiment Station, Chalk River, Ontario; Mary W. Ferguson, Thornhill, Ontario; Dr. R. Ogilvie, Curator of Botany, British Columbia Provincial Museum, Victoria; J. Kolynchuk, City Clerk, Saskatoon; Thomas K. Ovanin, Victoria; and Mary Charles and Sharon E. Godkin, Ottawa.

We should also like to thank all those who provided recipes and specific information on edible fruits: Constance Conrader, Oconomowoc, Wisconsin, whose recipes appeared originally in *Wisconsin Trails*; Eleanor A. Ellis, Department of Indian and Northern Affairs, Ottawa, whose recipes from *Northern Cookbook* are reproduced by permission of the Minister of Supply and Services, Ottawa; Dr. Dale Kinkade, Department of Linguistics, University of British Columbia, Vancouver; Enid K. Lemon and Lissa Calvert of Victoria, whose recipe is taken from their collection of printed notepaper, "Pick 'n' Cook Notes"; Dr. W. I. Illman, Department of Biology, Carleton University, Ottawa; Janet E. Renfroe, Victoria; Jane Chapman, Hornby Island, B.C.; and Marilyn Hirsekorn, Surrey, B.C. We are also pleased to be able to include recipes of the late Mary Szczawinski, wife of Adam Szczawinski, who collected fruit recipes for many years.

# Preface

Of all the different kinds of edible wild foods used by man, the wild fruits are without doubt the most widely enjoyed. Throughout Canada, berry-pickers and nut-gatherers venture out each year in search of their favourite local crops. Many also gather wild greens and other types of wild edibles, but wild fruits are generally easier to identify and often more palatable than other wild foods.

In pure economic terms, the gathering of wild fruits could be regarded as a wasteful and inefficient occupation, but who can possibly fix a dollar value on a dish of wild strawberry shortcake, a piece of fresh wild berry pie, or a spoonful of wild grape jelly on toast? An afternoon spent in the sunshine in search of the largest and sweetest wild blueberries the woods have to offer cannot be purchased with any currency known to man. No, there are more subtle values at stake than money. Perhaps foraging for wild crops satisfies some instinctive yearning left over from man's evolutionary past when this occupation was essential for survival. Whatever the reason, wild-fruit gatherers usually feel a real sense of accomplishment and fulfilment in collecting and using their harvest.

Having ourselves been harvesters of wild fruits for many years, we are deeply aware of the intangible values of this activity. However, utilizing wild fruits also has some very practical merit. We have both had the experience of being lost or stranded in the bush and being totally dependent on wild foods, mainly wild berries, for survival. Nutritionally, wild fruits are as rich in vitamins and minerals as cultivated fruits, and are usually lower in sugar and calorie content, not a disadvantage for most North Americans. Furthermore, wild fruits seem to be more flavourful than commercial fruits.

There are hundreds of different species of edible wild fruits in Canada, including not only the more popular berries and berry-like fruits, but also various nuts and lesser-known fruits such as grains and achenes. Some of these are easily prepared and delicious by any standards. Others are of more marginal utility, being difficult to gather or requiring special preparation to make them palatable, but as emergency foods they are important to know about.

Included in this book are over thirty-five of the most common and interesting edible wild fruit species or groups of species in Canada, with their botanical descriptions, notes on general habitat and distribution in this country, information on their collection and preparation, and specific recipes for each. Some of these recipes are based on those used by native Indian and Inuit peoples or by early pioneers, who were, of course, far more dependent on wild fruits than we are today. Others have been adapted to contemporary techniques of food preparation. Some can be prepared easily over a campstove or campfire; others call for the use of electrical appliances and a well-stocked kitchen.

Also provided, where possible, are notes of interest on the history and world-wide significance of the fruits and their relatives, and information on other applications of the plants in technology and medicine.

**Editor's Note**

The transition from the imperial system of weights and measures to the international metric system has prompted the use of both systems in this book. The recipes were originally developed under the imperial system, and have been converted to their closest metric equivalents based on the Canadian Metric Commission's guidelines. These "equivalents" are actually replacements, and sometimes involve slightly larger amounts than the original imperial measures. For example, the 8-ounce cup is actually equivalent to 225 mL, but the standard metric measure that replaces it holds 250 mL. As a result, the metric conversions of the recipes in this book sometimes yield slightly larger amounts than do the original recipes.

Readers of this series on edible wild plants will notice the change in the symbol for millilitres from ml in the first two publications, to mL in the present one. The change was prompted by the decision of the Metric Commission to adopt mL as the official symbol.

## What is a Fruit?

Few people would have difficulty identifying an apple, an orange, or a banana as a fruit, and most would agree that the different kinds of berries—strawberries, raspberries, blueberries—are also types of fruits. Botanically speaking, however, the term *fruit* encompasses a much broader category than most laymen realize. Nuts, of course, are fruits, but because they are rarely recognized as such, except in botanical terms, we have included specific reference to them in the title of this book.

A fruit is defined as "a ripened ovary, together with any other floral structures that ripen with it and form a unit with it". (The ovary is that part of the female organ in flowering plants that encloses the ovules, or young, undeveloped seeds.) By this definition, many of our so-called vegetables, including green beans, peppers, tomatoes, and squash, are actually fruits. Grains of grasses, nuts as mentioned above, and sunflower "seeds" when the outer shell is included are also fruits.

There are several ways of classifying fruits. One primary division is between those that are fleshy and those that are hard and dry. Fleshy fruits include true berries, drupes, and pomes. *Berries* have a thin outer skin surrounding a juicy pulp through which are distributed several to many small seeds. Some wild berries are gooseberries, currants, cranberries, blueberries, and grapes. The garden tomato is also a berry by definition. *Drupes* closely resemble berries, but the seeds, usually only one or two, are enclosed in a hard, bony, inner ovary wall, surrounded by a soft, but not always edible, outer layer. Well-known drupes are the cherry and plum. Others include elderberry, kinnikinnick, bearberry, crowberry, bunchberry, high-bush cranberry, and sumac. *Pomes*, of which the most common examples are the apple and pear, consist of a fleshy outer layer surrounding a core of bony or cartilaginous membranes that enclose the seeds. The soft portion consists, in part, of the ripened receptacle and calyx of the flower as well as the ovary. Pomes include crabapples and a number of other wild fruits related to the apple: saskatoon berries, mountain-ash fruits, and hawthorn berries.

The berry, drupe, and pome are all simple fleshy fruits. Some fruits, such as raspberries and blackberries, are known as *aggregate fruits*, developing from flowers with multiple pistils that ripen together into a mass of tiny fruits, or drupelets. Strawberries are also aggregate fruits, with the main fleshy portion being the ripened receptacle; the tiny encased seeds are borne on the surface of the fruit. Mulberries resemble blackberries in appearance but are classed as *multiple fruits*, because they are derived from a number of separate flowers, each with a single pistil, rather than from a single flower with many pistils.

The fruits of soapberry, buffaloberry, and salal, while berry-like, do not fit into any of the other fleshy fruit categories. They are known as *accessory fruits*, since the main fleshy part is derived not from the ovary but from the thickened calyx of the flower. Rose fruits, known as *hips*, are also unique, consisting of a fleshy receptacle, contracted at the mouth, that encloses a mass of bony "seeds" (achenes), each derived from a single pistil.

The dry, hard fruits are classified into those that at maturity split open along definite seams and those that do not. Those that do include some well-known edible types, such as beans and peas, but most are not edible. There are no true examples in this book of fruits in this category, with the possible exception of yellow pond-lily, whose fruit is classed as a *capsule*, but the pulp, unlike that of most capsules, does not dry out, and the fruit splits open irregularly when mature.

Fruits that do not split open at maturity include achenes, grains, and nuts. *Achenes*, such as those of the sunflower and the balsamroot, are small, unwinged, one-seeded fruits in which the seed separates from its outer covering. Grains, including wheat, corn, oats, rice, and the various wild grasses, are similar to achenes except that the outer covering is tightly fused to the seed kernel.

*Nuts* are a more difficult category because the common conception of a nut is somewhat different from the botanical definition. An almond, for example, is actually a drupe, like a plum, but it is the seed kernel, not the outer flesh, that is the edible portion. Walnuts are also drupe-like but do not fit well into either the fleshy or dry fruit categories. Lotus-lily fruits are nut-like but are not classed as true nuts because of the unique receptacle enclosing them. Peanuts, also, are not true nuts, but with peas and beans belong to the first category of dry fruits – those that split open. True nuts are wholly or partially enclosed in a papery, woody, or spiny husk, and the nut itself has a bony or leathery outer wall containing usually only one seed. Examples include acorns (oak fruits), hazelnuts, and hickories.

Of course, not all fruits are edible; some are too rough and dry, some too small to bother with, some bad tasting, and some poisonous. There are no set rules for determining whether a fruit is harmful or not. Some poisonous fruits look very inviting, while many edible ones do not seem particularly promising. It is always best to stay away from any berry or other fruit unless you recognize it and are positive it is edible.

This book includes some of the most common and best-liked of Canada's wild fruits. There are many others that are too restricted in range, barely edible, or too easily confused with poisonous species to include. It should also be mentioned that almost all the fruits described here are native to Canada. In addition to the indigenous species discussed, there are a number of introduced fruits, such as apples, plums, pears, cherries, grapes, and blackberries, that have escaped cultivation or persisted in old orchards and homesteads. These domestic fruits "gone wild" are usually smaller and tarter than those under cultivation, but are often tastier and more suitable for cooking.

The fruits discussed here vary considerably in quality and potential. Some of the tarter fruits, such as Oregon grapes, highbush cranberries, buffaloberries, and wild crabapples, are at their best when used to make jelly or when mixed with blander fruits. Many of the fruits are excellent for baking in cookies, muffins, and pies. With some fruits, however, notably wild strawberries and arctic raspberries, there is no method of preparation that could in any way improve on the plain, fresh berries, picked by the handful and popped directly into the mouth.

## The Value of Wild Fruits

Have you ever wondered why, in this modern age, people still like to gather wild fruits? Why do they willingly spend all the hours necessary for picking and preparing them when they can buy commercial equivalents at the nearest supermarket? The answer is obvious to those who follow this seemingly illogical pursuit. Firstly, wild fruits are free for the taking. Although the harvester may have spent a considerable amount of time and even money getting to where the wild fruits are, and, of course, must expend time and energy in the harvest, the product itself is free and the pleasures of the outing are a bonus. Furthermore, wild fruits can add greatly to the variety in our diet. To those who like adventure in cuisine, there is nothing better than some new berry to add to a dessert, or some wild nuts or seeds to cook in a gourmet soup. When we use these wild products, especially when camping or serving a completely wild dinner, it gives a special sense of history: these were the foods of the first Americans— the Indians and the Inuit—and of the early European settlers and explorers.

In addition, many of the wild fruits, such as cranberries, blueberries, strawberries, raspberries, blackberries, and grapes, are direct ancestors of the cultivated forms that have been developed through years of careful crossing and selection. Eating these wild types is like opening a window onto the past; one has to admire the men and women of previous generations who had the foresight to recognize their potential as commercial crops. We would do well to follow the example of these horticultural pioneers and realize the potential of some other wild fruits. Saskatoon berries, for example, have often been considered as a possible commercial crop, and experiments have been under way for some time to propagate and grow them on a large scale. Several excellent varieties have been recognized, including one in Alberta, known as "smoky", that is already being used commercially in some areas, and another in Saskatchewan, called "honeywood", that shows considerable promise.

Salal fruits, so prolific on the Pacific Coast, should also be carefully considered for their commercial potential, as should more of the wild blueberry and huckleberry species. High-bush cranberries, elderberries, Oregon grapes, and some of the nut trees are already being grown for their ornamental value, but they should be considered for their value as crop species as well.

The nutritional value of some wild fruits is already well known. Rose hips are famous as a source of vitamin C: one study found rose hips to contain about thirty times by weight the vitamin C in orange juice. Many other wild fruits contain significant quantities of vitamin C—cranberries, high-bush cranberries, crowberries, bearberries, currants, and cloudberries, to name just a few.

Although the record for other vitamins and minerals is incomplete, available information indicates that the wild fruits as a group are equal nutritionally to the commercial types, and in many cases surpass them. Elderberries, for example, contain almost three times the vitamin A found in peaches, themselves known to be high in that vitamin. Elderberries are also very high in calcium, phosphorus, and iron. Saskatoon berries are among the most iron-rich fruits known to man. In one study, dried saskatoons were found to contain over twice the iron found in dried prunes and over three times that in raisins.

Wild fruits, especially the nuts and nut-like fruits, also contain proteins and oils, and all fruits contain carbohydrates. The value of fruits as survival foods should not be underestimated. Some, such as high-bush cranberries, low-bush cranberries, rose hips, kinnikinnick, bearberries, and crowberries can be found on the plants even in midwinter, and almost all can be preserved for winter by drying, canning, or freezing. Indians and Inuit were masters at utilizing wild fruits on a year-round basis, for these fruits were not a luxury to them but a major means of survival.

## Gathering and Preparing Wild Fruits

The careful harvesting of wild fruits is not at all destructive to the plants. In fact, the very function of fruits—to promote the dispersal of the enclosed seeds and extend the range of the species—is often unwittingly served by human harvesters, as it is by birds and other animals.

Wild fruits are the easiest of all wild edibles to gather. In most cases all that is needed is a pail and nimble fingers. We recommend that you tie a long string to your container and loop it over your neck; in this way both hands are freed for picking. Indian women often strapped their picking containers around their waists. If you plan to gather large quantities of fruits, carry two or three containers of different sizes. You can pick into the smaller one, and empty it into the larger ones as it fills. Try not to overfill your containers, or the fruits lower down may be crushed and bruised.

Many fruits, especially the berries and berry-like fruits, are ideal for eating when raw and fresh. If you do not plan to use them immediately after harvesting, store the ripe ones in a cool, dark place (ideally a refrigerator), and leave the rest to ripen at room temperature for awhile. They will be less likely to bruise and spoil if you do not wash or handle them until just before using them.

Some fruits can even be gathered in winter and spring, but they inevitably lose some of their flavour and nutritive value. Except in emergencies, it is better to collect fruits at their peak of ripeness and, if you want them year-round, to preserve them by drying, freezing, or canning, or by making jams, jellies, syrups, juices, or wines. During the time you are actually preparing the fruit, the addition of a small amount of lemon juice, vinegar, or ascorbic acid (vitamin C) will help prevent discolouration and loss of flavour.

Drying is the simplest and oldest method of preserving most fruits. It can be done without the use of electricity or any special appliances and without any ingredients other than the fruits themselves. All one needs is a clean mat or screen and a warm, dry, well-ventilated spot. The drying process may take several days, and the main problem is to prevent the fruit from moulding or spoiling before it is dry, but when thoroughly dehydrated the fruit will keep all winter and even for several years if stored properly. It can be used in a dried state or can be rehydrated at any time by soaking in water.

Today, food-drying is regaining some of its former popularity and there are a number of compact electrical food dehydrators on the market. These have the advantage of being fast, independent of weather conditions, and hygienic. (Open-air fruit-drying often attracts flies and wasps, which can be a problem unless the fruit is covered with screening.) It is relatively simple to construct your own dehydrator, using frames with fibreglass screening, lightbulbs as a heat source, and a small fan to circulate the air.

Other methods of preserving wild fruits, such as freezing, canning, and making jams and jellies, are more familiar to most people today than drying. The techniques are described in many cookbooks, and several government publications on fruit-preserving are available to the public at no cost. These publications and others on fruit preparation are listed in the bibliography at the end of this book.

Wine-making is an art unto itself, but, as with jam- and jelly-making, the basic techniques are the same for almost all fruits. With a little experimentation, you should be able to make wines from most of the fleshy fruits in this book.

Wild nuts and seeds require little preparation, other than removing the outer husks and shells. Preserving these fruits is usually as simple as drying them slightly and storing in a cool, dry place. It is best to leave the shells on until just before they are to be used.

## A Necessary Caution

As has already been mentioned, not all fruits are edible: many are unpalatable, and a small number are poisonous. Furthermore, that the fruits of a plant are edible is no indication that other parts of the plant can be eaten. Quite the opposite is true in a number of cases: for example, the foliage, rhizomes, and seeds of the may-apple can cause severe illness and even death when consumed. Choke cherry leaves and stones contain cyanide and should never be eaten. Moreover, some berries, such as mulberry and red elderberry, can be toxic when unripe. Some fruits are quite edible when properly prepared, but can be poisonous in a raw or untreated state. For example, red elderberries are suspected of causing stomach upset and nausea when eaten raw, although cooked they were an important food for some Indian peoples. Acorns, a staple food of many native groups in North America, must, in the red oak species, be leached by boiling or soaking in water to remove the bitter, poisonous tannins. The foliage of both elderberries and oaks is also poisonous.

The reader is advised to pay special heed to the Warning notes provided in the discussions of some species. These are included whenever there is a possibility that the fruit or plant in question might be toxic or could be confused with other potentially harmful species. We discuss chemicals that may cause problems if consumed in large amounts, for example oxalic acid, nitrates, tannins, saponins, and cyanogenic glycosides. The presence of some of these chemicals has been the basis for the use of these plants in herbal medicines for centuries, and obviously one should not consume any medicinal preparations in quantity. These chemicals are also found in many ordinary garden vegetables and fruits, for which we could compile a similar list of warnings. Our bodies are quite adept at handling small amounts of chemicals, and as with the familiar garden vegetables and fruits we realize that the best diet consists of a variety of foods in moderation. In this way our bodies will obtain the necessary nutrients, which vary between plants, and will be able to handle the small amounts of toxic chemicals. We hope that this information will put into perspective the warnings found in this book.

In reviewing the toxicity of the various plants we were handicapped by having little information on the actual quantity of the particular chemicals in the plant or plant parts, and the dearth of adequate studies on the toxicity, including carcinogenicity, of most plant chemicals. We realize that the quantities of chemicals in plants may vary depending on plant variety, soil, climatic conditions, and the maturity of the plant. For these reasons we cannot state the quantity of a food that might safely be consumed: we can only urge that the reader use common sense. Much of our information on safety comes from personal experience with these plants. Thus if you make a delicious hawthorn jelly, don't get carried away—use it with moderation.

Further information on chemicals in plants, including house plants and garden vegetables, can be found in John M. Kingsbury's *Deadly Harvest: A Guide to Common Poisonous Plants.*

Wild fruits have the advantage over commercial ones of usually being free from the insecticides and other harmful chemicals used on many agricultural crops. However, some pesticides are also applied over forested areas for control of mosquitoes and tree-damaging insects. Furthermore, fruits gathered along roadsides, railway rights-of-way, edges of cultivated fields, or power-line clearings may be contaminated with herbicides (weed-killers). The wild-fruit gatherer should be careful to avoid picking in areas that have been sprayed or in any place where the plants themselves do not look healthy. As an extra precaution, the fruits should always be washed in clear, cold water before use.

## Format

This book includes, by the very definition of the term *fruit*, only flowering plants (angiosperms). With the exception of wild rice and other wild grains (monocotyledonous plants in the grass family—Poaceae or Gramineae), all are dicotyledonous flowering plants. The wild grains are discussed first, and the others, in the subdivision Dicotyledonae, are arranged in alphabetical order by scientific family name. When more than one species is described within a family, they are listed in alphabetical order of the plants' scientific names (genus and species).

The scientific names are included for your convenience, to ensure that there is no doubt about the particular species being discussed. Because some plants or groups of plants may have two, three, or more different colloquial, or common, names in various parts of the country, and because the same common name may be applied by different people to entirely different species, scientific names are both useful and necessary in distinguishing the exact identity of a plant. Scientific names are also necessary when two or more languages are involved; they are standardized the world over and are thus recognized by speakers of any language.

For each species, the most commonly applied colloquial names of both the plant and its family are included in the top outer margin. The corresponding scientific names are given at the bottom of the page. When two or more related species can be used similarly, they are treated in the same section. Alternative or localized common names are given in the text.

We have tried, where possible, to avoid the use of technical terms or botanical jargon, but sometimes it is necessary to use specialized words that might be unfamiliar. These words are defined in a glossary towards the end of the book.

# Edible
# Wild Fruits
# and Nuts

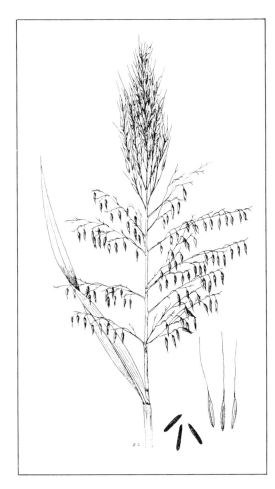

## Other Names
Wild rice is also called Indian rice, water rice, water oats, and Tuscarora rice.

## How to Recognize
Few people would consider including wild grains in a discussion on edible wild fruits but, although they are not fleshy, grains are as much a fruit by definition as wild crabapples, blackberries, or strawberries. Botanically, a grain, or grass fruit, is known as a *caryopsis.* It consists of a ripened ovary fused to an outer seed coat.

Grains are the most valuable of all sources of plant food for man, both directly as cereals, flour, and meal, and indirectly as fodder for livestock that provide dairy products and meat. Grains contain mostly starch, with some proteins and traces of vitamins and minerals. When grains are mentioned, most of us think of such domesticated crops as wheat, rye, oats, barley, corn, and rice, but there are many wild species that can be utilized in the same way.

# Wild Rice and Other Wild Grains

(Grass Family)

***Zizania aquatica* L.
and other grass species**
(Poaceae or Gramineae)

In Canada the most famous of the wild grains is wild rice. Used for centuries by Indians, it has now become a modest commercial crop and is sold as a gourmet item throughout North America. Of the other wild grains, most are too scattered or have fruits too small to be considered as a practical food source. Also, the grains of many are too tightly enclosed in the chaff, and can be separated only by parching or soaking in lye. However, a number of native species, including manna grass (*Glyceria* species), reed-grass (*Phragmites communis* L.), rye grass (*Elymus* species), drop-seed grass (*Sporobolus* species), rice grass (*Oryzopsis* species), slough grass [*Beckmannia syzigachne* (Steud.) Fern.], and canary reed-grass (*Phalaris* species), do have grains that are harvestable and suitable for use as food. In addition, there are several edible weedy or introduced species, such as barnyard grass [*Echinochloa crusgalli* (L.) Beauv.], as described in our earlier publication, *Edible Garden Weeds of Canada*.

Wild rice is a robust annual grass up to 3 m or more high. The leaves are broad and soft, and the stems quite stout and occasionally branched. The flower clusters are long and broom-like, with the spreading male, or pollen-bearing, flowers below, and the erect female, or seed-bearing, flowers near the top. The fruits, 14 to 20 per head, are awl-shaped, cylindrical, and often over 1 cm long. They are black and shiny and are enclosed in a loose, bristle-tipped husk. They ripen from early summer to autumn, dropping off quickly after maturity.

**Where to Find**

Wild rice is found along watercourses, marshes, and lake-margins from eastern Quebec and New Brunswick to Manitoba. Most of the other grasses mentioned are widespread, and are common in at least some regions of Canada. A number, including manna grass, reed-grass, slough grass, and canary reed-grass, grow in wetland habitats. Others, such as the rye grass species, grow in a variety of habitats, from seashores to open woodlands. Drop-seed grass and rice grass tend to prefer drier, upland sites.

## How to Use

To native peoples in many parts of eastern North America, particularly in the Great Lakes region, wild rice has been a staple food. The Menomini Indians of Wisconsin are famous for their use of wild rice, and the tribal name is actually derived from their native word for rice. If you do collect wild rice, be sure that you do not trespass on lands where Indian people have harvesting rights, or on other private property, as this valuable grain is still being gathered for private and commercial use. The best method of harvesting and preparing the grain is in the time-honoured ways of the native people. Since it grows in standing water, it must be gathered by boat or canoe: one person guides the boat through the wild rice stand, and another bends the fruiting heads into the boat and shakes or raps them sharply to dislodge the grains. When enough has been gathered, the kernels must be cleaned and the awns (bristles) and hulls removed by first heating them in a dry kettle or skillet, then beating or trampling the grains.

The chaff is removed by winnowing. When the grains are tossed into the wind or in front of a fan, the light chaff is blown away and the heavier grains drop back onto the tray. Next, the grain should be washed, and after being dried is ready to cook or store.

Wild rice can be purchased in specialty shops or in the gourmet sections of some grocery stores. It is expensive, but considering the work required to gather and prepare it, the price is not unreasonable. Commercial wild rice needs little preparation, other than washing, before being cooked.

The grains of wild rice are much longer than those of conventional rice, and when cooked they split and curl back at the ends. Wild rice can be mixed with regular brown rice and is delicious when cooked with mushrooms, or with almonds or other nuts. The grains are firmer than those of ordinary rice and have a pleasant, nutty taste.

The other wild grains are much smaller, but their preparation is very similar. In some, such as reed-grass, the hulls cannot be removed, even with parching, and must be cooked and eaten with the grains. Wild grains can be dried, ground into meal, and used in baking muffins, breads, cakes, and in thickening soups. They can even be dark-roasted, ground to a powder, and used to make a coffee substitute. In an emergency, wild grains can be cooked with water to make a nourishing mush.

## Warning

Care should be taken in gathering wild grains to avoid those infected with the extremely poisonous fungus known as ergot (*Claviceps* species), which appears in place of the regular grain as a small, hard, black kernel. Ergot can also contaminate domesticated grains, and in the past has caused sickness and death in both humans and livestock. When winnowing the chaff of wild grains, be sure to remove all the outer bristles, or awns; they could become lodged in the throat and cause choking.

## Suggested Recipes

### Plain Wild Rice

| | |
|---|---|
| 250 mL  wild rice | 1 cup |
| 750 mL  boiling water | 3 cups |
| 5 mL  salt | 1 tsp |

Wash wild rice well in cold water, drain, and stir slowly into the boiling, salted water. Cook without stirring until tender, about 40 minutes. Alternatively, the rice can be parboiled about 5 minutes, then removed from the heat and left to soak, covered, for about 1 hour. Other wild grains can be cooked in the same way, but the ratio of water to grain will vary. Serves 3–4.

For an interesting variation of this basic recipe, substitute deer (or pork) broth for the water, and sprinkle with maple sugar before serving. This recipe was adapted from a description of the Menomini Indian way of serving wild rice in *Ethnobotany of the Menomini Indians* by Huron H. Smith.

## Wild Rice Stuffed Peppers

| | | |
|---|---|---|
| 250 mL | uncooked wild rice | 1 cup |
| 750 mL | boiling water | 3 cups |
| 1 mL | salt | 1/4 tsp |
| 50 mL | butter *or* margarine | 1/4 cup |
| | 1 medium onion, chopped | |
| 500 mL | sliced mushrooms | 2 cups |
| 5 mL | sweet basil | 1 tsp |
| 5 mL | oregano | 1 tsp |
| 5 mL | parsley | 1 tsp |
| 250 mL | grated sharp cheddar cheese | 1 cup |
| 500 mL | canned tomatoes | 2 cups |
| | 8 large, sweet green peppers | |

Wash wild rice in cold water, drain, and add to boiling, salted water. Parboil 5 minutes, then remove from heat and let soak about 1 hour. Drain. Sauté onion and mushrooms in butter for about 5 minutes. Add cooked rice and all remaining ingredients except the peppers. Wash peppers, carefully cut around and remove tops, leaving them intact, and cut out seeds and core. Spoon wild rice mixture into pepper cavities, packing in well. Then replace tops. Place peppers upright in a deep casserole, add a small amount (about 2 cm) of water to the bottom. Bake at 180°C (350°F) about 1 hour, or until peppers are soft but not mushy. Serve hot. Serves 8.

## Wild Rice Dressing for Game

| | | |
|---|---|---|
| 250 mL | uncooked wild rice | 1 cup |
| 750 mL | water *or* giblet stock | 3 cups |
| 5 mL | salt | 1 tsp |
| 50 mL | butter | 1/4 cup |
| 25 mL | green onions, chopped | 1 1/2 tbsp |
| 30 mL | green peppers, chopped | 2 tbsp |
| 125 mL | celery, chopped | 1/2 cup |
| 125 mL | mushrooms, sliced | 1/2 cup |
| 125 mL | almonds, chopped | 1/2 cup |
| 50 mL | tomato paste | 1/4 cup |
| 5 mL | fresh sage | 1 tsp |
| | *or* | |
| 1 mL | powdered sage | 1/4 tsp |
| 5 mL | parsley | 1 tsp |
| | dash of garlic salt (optional) | |

Wash wild rice. Bring water *or* stock to boiling point and add salt and wild rice. Simmer about 30 minutes or until rice is tender but not mushy. Drain off any excess liquid. Melt butter in a skillet and gently sauté onions, peppers, celery, mushrooms, and almonds. Add the hot, drained rice, tomato paste, and seasonings and mix well. Use as stuffing for wild game or poultry, or bake separately and serve with game. Giblets, cooked and finely chopped, may be added to the dressing. Do not stuff game or poultry until just prior to cooking, and serve immediately when cooked.

## Wild Rice Griddle Cakes

|  |  |  |
|---|---|---|
|  | 3 eggs, well beaten |  |
| 500 mL | buttermilk | 2 cups |
| 15 mL | honey | 1 tbsp |
| 500 mL | whole-wheat flour | 2 cups |
| 10 mL | baking powder | 2 tsp |
| 5 mL | baking soda | 1 tsp |
| 5 mL | salt | 1 tsp |
| 25 mL | vegetable oil | 1½ tbsp |
| 250 mL | cooked wild rice | 1 cup |

Mix together eggs, buttermilk, and honey. Sift together dry ingredients and gradually add to the liquid, beating until smooth after each addition. Stir in oil and cooked rice and drop by large spoonfuls onto a hot, greased griddle. Cook as for any pancakes. Serve immediately with sausages and maple syrup. Serves 4–6.

## More for Your Interest

Wild rice has been an important food source not only for humans but for waterfowl of all kinds. The following account by early fur trader Peter Pond of a journey in Wisconsin in the last part of the eighteenth century is a good illustration: "We Came to a Shallo Lake whare you Could Sea no water But Just in the Caneu track[.] the Wild Oates [wild rice] was so thick that the Indans Could Scarse Git one of thare Small Canues into it to Geather it and the Wild Ducks Whe[n] thay Ris Maed a nois like thund[er]" (Gates, ed., 1965, p. 36). In many areas of eastern North America, wild rice has been planted around lakes and along waterways solely to attract ducks and geese.

Among the Menomini Indians it was said that in favourable sites one man could reap as much as 20 kg of wild rice in one day, and that a good season's harvest was about 400 kg.

Wild rice is very nutritious, being low in fat, high in protein, and rich in vitamin B. It compares favourably with cultivated grains, such as wheat, corn, and rye, in thiamine content, and is richer in riboflavin than wheat, corn, oats, or rye.

# Staghorn Sumac and Smooth Sumac

(Cashew Family)

## Other Names
Sumac, sumac-tree, vinegar-tree.

## How to Recognize
These are deciduous shrubs, or sometimes (in the case of staghorn sumac) small trees, that divide near the ground into thick, upright branches. *R. typhina* grows up to 10 m tall, and *R. glabra* is usually under 3 m but occasionally grows as high as 5 m. When broken, the young twigs exude a thick, milky juice. The leaves are large and pinnately compound, with 11 to 29 or so individual leaflets in opposite pairs, with one terminal one. The leaflets are narrowly elliptical or lance-shaped, and sharply pointed at the tips, with finely toothed edges. The middle pairs of leaflets are longer than those at either end of the leaf. The leaves, dark green above and paler beneath, turn a brilliant orange or scarlet in the fall. The yellowish-green flowers of these species are borne in dense, upright clusters at the ends of the branches, male and female usually on separate plants. The sticky, sour-tasting fruits are dense, cone-shaped "candles" of hard, single-seeded, bright-red to deep-scarlet drupes, and usually remain on the branches over the winter.

The twigs, leafstalks, and fruits of the staghorn sumac are densely fuzzy, whereas in smooth sumac they are relatively smooth. These species have been known to hybridize where their ranges overlap.

There are several other *Rhus* species in Canada, including the well-known poison ivy (*R. radicans* L.), poison oak (*R. diversiloba* T. & G.), and poison sumac (*R. vernix* L.). These species contain a milky or clear, slightly volatile oil to which many people are allergic. It causes a severe burning or itching, accompanied by a blistering rash, to those who are sensitive to it. Fortunately these species are easy to distinguish from the staghorn and smooth sumacs: poison ivy and poison oak are low shrubs or creeping vines with 3-parted leaves; poison sumac is a tree with 7- to 13-parted leaves with white fruits in relatively small, open clusters.

*Rhus typhina* L. and *R. glabra* L.
(Anacardiaceae)

## Where to Find

Staghorn and smooth sumacs usually grow in open areas in sandy or rocky soils. Staghorn sumac is found from southeastern Ontario east to Nova Scotia. Smooth sumac ranges from the dry interior of British Columbia east to Lake Huron and in the United States to New England and Georgia.

## How to Use

The hard, sour fruits of the staghorn and smooth sumacs are not particularly good to eat raw, although they have a tangy flavour and some may find them pleasant to nibble on during outings. Sumac juice makes an appetizing jelly, especially good with meats. Dr. Anne Marie Stewart and Leon Kronoff, in their interesting book *Eating from the Wild*, have even devised recipes for "sumac borscht" and "sumac cream sauce", which we recommend.

The best time to pick the fruits for juice or jelly is in midsummer, before rain, when their colouring is at its brightest and the tiny

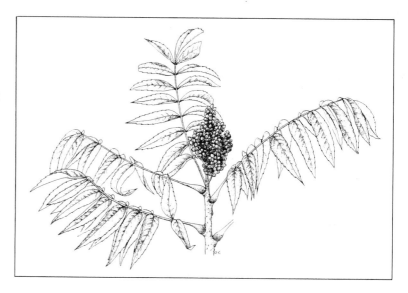

hairs on the fruit are still intact. When the rain washes off the malic acid contained in the hairs, the fruit loses its fresh flavour. Sumac fruits can also be harvested in fall or even winter, but their quality deteriorates somewhat with age. It is easy to gather a large quantity of the fruits, as the tight clusters snap off the branches easily. To prepare the fruits for cooking simply break off the small clusters, discarding the main central stem. There is no need to separate the fruits individually. They can be stored for several weeks after harvesting and can be frozen for later use.

### Warning

Sumac contains high quantities of tannic acid (see Warning in oaks section, p. 102). Use in moderation, and avoid prolonged boiling as this extracts more of the tannic acid.

### Suggested Recipes

### Sumac-ade

|  | 6–8 medium-sized sumac fruit clusters | |
|---|---|---|
| 1.5 L | water | 6 cups |
| 125 mL | sugar *or* maple sugar *or* honey | ½ cup |

Wash the sumac fruits and break them off the main stems of the clusters. Place with the water in a large bowl or saucepan, and bruise the fruits thoroughly until the water becomes a light-pink colour. You can heat the mixture until just before boiling to extract more of the juice. (However, the high concentration of tannic acid found in sumac fruits is released when they are cooked too long and will make the juice bitter; see Warning above). Remove the large pieces of sumac and pour the liquid through a jelly bag or fine sieve to strain out the smaller pieces and hairs from the fruits. Add sugar or other sweetening and stir until dissolved. Cool and serve with ice cubes on a hot summer day. Makes about 12 small glasses. This beverage is also pleasant when mixed half and half with lemonade or with other fruit juices.

## Sumac–Blue-Plum Jelly

| | 12 sumac fruit clusters | |
|---|---|---|
| 1 L | water | 4 cups |
| | blue-plum juice | |
| 1.25 L | sugar | 5 cups |
| 57 g | pectin crystals (1 package) | 2 oz |
| 15 mL | lemon juice | 1 tbsp |

Clean sumac clusters and break into small pieces, removing the central stems. Simmer in water until the juice is a dark-pink colour (as short a time as possible to avoid extracting too much tannic acid, which will make the juice bitter). Pour through a jelly bag and allow to drain for several hours, or until dripping has stopped. Measure the juice and combine 1 L (4 cups) of juice with 500 mL (2 cups) blue-plum juice (made by boiling and straining plums); add pectin crystals and lemon juice. Heat and stir until mixture comes to a full boil. Add sugar, bring to a boil again, and boil hard for 1 minute, stirring constantly. Then pour into hot, sterilized jelly glasses, seal and label. Store in a cool place. Yields about 8 medium-sized jelly glasses. (Courtesy of Dr. W. I. Illman.)

## More for Your Interest

The sumacs are in the same family as the popular cashew nut and the mango. The Iroquois, the Menomini Indians, and others made a beverage from sumac fruits (but see Warning above). The Lillooet Indians of the British Columbia interior used the white, milky juice from the broken twigs to remove warts. Some Indian peoples used the leaves in smoking mixtures.

Because of their brightly coloured, long-lasting fruits and their brilliant fall foliage, the staghorn and smooth sumacs are popular as ornamental plants, and can be found growing in gardens throughout southern Canada. A form of the staghorn sumac with finely cut leaves (*R. typhina* f. *dissecta* Rehder) is commonly seen in cultivation. Sumacs spread widely by long, shallow roots and tend to form large thickets. For this reason they can be a nuisance in some gardens and should be planted only in areas where they have plenty of space.

Sumac wood is light, soft, and yellowish or orange. It is sometimes used for decorative finishing and for novelties.

## How to Recognize

This plant is a large annual, often exceeding 2 m in height, with rough, hairy stems and leaves. The leaves are broadly oval, often heart-shaped at the base, with long stems and toothed edges. The flower heads, which always turn towards the sun, following it through the sky, are large with a broad, flat, brownish or yellowish disk and bright-yellow petal-like rays, characteristic of the sunflowers and their relatives. The common wild sunflower is the ancestor of our cultivated sunflower. The flower heads are smaller, and the fruits, botanically known as achenes, are smaller and less numerous than those of the cultivated form, but otherwise there is little difference between them. If left without proper cultivation, the horticultural sunflower will quickly revert to its wild form. Several other species of wild sunflower occur in North America, all with edible seeds.

Another plant in the sunflower group, known as balsamroot or spring sunflower [*Balsamorhiza sagittata* (Pursh) Nutt.], also has edible achenes. It is a low perennial up to half a metre high, with numerous silvery arrowhead-shaped leaves growing in a clump from a thick rootstock. The flower heads have showy yellow rays like those of the sunflower. The achenes are much smaller than those of the sunflower, but are similar in appearance.

## Where to Find

The common wild sunflower is native to western North America, but has been widely established through Indian cultivation and now occurs as a weed throughout much of North America and in Europe as well. It and other sunflower species can be found in many parts of southern Canada from British Columbia to the Maritimes, growing in fields, pastures, and along roadsides and river margins. Balsamroot is restricted to the dry open hillsides and flatlands in the southern interior of British Columbia and the foothills of Alberta.

# Common Wild Sunflower

(Aster or Composite Family)

*Helianthus annuus* L.
(Asteraceae or Compositae)

## How to Use

The seeds of wild sunflower, though smaller than those of the cultivated sunflower, are very numerous and easy to gather in large quantities: simply shake the fully mature fruiting heads into a tray or bowl. Like commercial sunflower seeds they are rich in protein, oil, minerals, and vitamins, especially vitamin D, which is not available in most foods.

Wild sunflower seeds may be eaten fresh, parched and ground into meal, or pounded and gently boiled to extract the oil. The shells are somewhat of an inconvenience, but if the little fruits are gathered while still soft and succulent, the shells can be ground up along with the kernels. The green fruits have the same pleasant, nutty aroma as the flower heads. Ripe sunflower seeds can be roasted, ground and used alone, or mixed with roasted barley as a very acceptable substitute for coffee. Wild sunflower seeds can, of course, be substituted for the commercial ones called for in recipes if you have the time and patience to extract the kernels. However, we suggest that they are best used in the more traditional ways of the North American Indians, and the recipes we include were inspired by native cookery.

Balsamroot seeds are even smaller than those of wild sunflower, and, with them, removing the outer shells is out of the question. The outer covering is simply ground up with the kernel into meal, which can be eaten alone, mixed with berries or dried meat, or used as a thickening agent for soups and stews.

## Suggested Recipes

## Sunflower Seed Meal

Dry sunflower seeds thoroughly by spreading them out in a well-ventilated place for a few days. Parch the seeds in a hot skillet until well browned, then grind them in a food mill or blender until they are of powder-like consistency. If the seeds are harvested while still green and tender, the shells may be ground up with the kernels. Otherwise they must be removed after the seeds are dried and parched.

## Sunflower–Cornmeal Cakes

| | | |
|---|---|---|
| 500 mL | sunflower seed meal | 2 cups |
| 10 mL | salt | 2 tsp |
| 500 mL | water | 2 cups |
| 125 mL | cornmeal | ½ cup |
| 50 mL | vegetable oil | ¼ cup |

Mix together sunflower seed meal and salt in a saucepan. Add enough water to make a thick, sauce-like consistency (you may need more than 500 mL, depending on the fineness of your meal). Cover and simmer for half an hour, stirring occasionally. Mix in sufficient cornmeal to make a dough stiff enough to be shaped with the hands. Cool, then shape into round, flat cakes about the size of tea biscuits. Heat oil in a large, heavy skillet until it starts to sizzle. Drop the cakes into the oil, cook until well browned, then turn and brown the other side. Drain on paper towelling and serve hot or cold with butter or honey. Makes about 12 cakes.

## Sunflower Seed Oil

| | | |
|---|---|---|
| 500 mL | unshelled sunflower seeds | 2 cups |
| 500 mL | water | 2 cups |

Crush the sunflower seeds to a pulp with mortar and pestle or with a rolling pin on a board. Add the pulp to the water and bring to a boil. Reduce heat and simmer gently for about 1 hour. Strain off solid particles and cool, then skim the oil off the top of the water. This makes an excellent table oil; it was used by the Indians for annointing the skin and hair. The above quantity of seeds should yield about 50 mL (¼ cup) of oil.

## More for Your Interest

Originating in western North America or Mexico, the sunflower was cultivated by the Indians of that region many centuries ago, and its use gradually spread to the Indians of the Plains and eastern North America. In many areas the seeds became a staple food. In 1615, Samuel de Champlain noted the culture of the sunflower among the Huron people of Georgian Bay and Lake Simcoe. They grew it in with their maize and used the kernels both for oil and as a meal for mixing in maize soup. As early as the mid-sixteenth century the sunflower was introduced to Europe and became widely cultivated, especially in eastern Europe and Russia.

Today the sunflower is grown in many temperate and tropical countries of the world, including Canada and Argentina. Sunflower oil is used for cooking and in salads, as well as in making margarine, varnishes, and soaps. The residue left after the oil is extracted is valuable as cattle food, and the stems and leaves are also an excellent fodder. In Germany the dried leaves are used locally as a tobacco substitute. In addition, the seed receptacles are made into blotting paper, the inner stalk into fine writing paper, and the stem fibres into a silky, flax-like thread. The flowers are attractive to honeybees, and the flower buds and young stems make an acceptable green vegetable. The seeds themselves are eaten not only by humans but by poultry and birds of all kinds, both wild and domesticated. Surely there are few plants more versatile.

## Other Names
Mahonia, mountain grape, or wild barberry; *B. aquifolium* is sometimes called tall Oregon grape or tall mahonia.

## How to Recognize
These are low, straggling to erect evergreen shrubs, usually under 1 m tall. The leaves are leathery and compound, the individual segments resembling holly leaves, being shiny with prickly margins. The bark is greyish outside, bright yellow inside. The bright-yellow flowers, in conspicuous elongated clusters, appear in early spring. When ripe the berries are round, smooth, and deep blue, with a whitish waxy coating. They are juicy but very tart and quite seedy. Of the two species, *B. aquifolium* is taller and has 5 to 7 leaflets per leaf, whereas *B. nervosa* usually has 9 to 15 leaflets. These species are classed by some botanists in the genus *Mahonia*. A closely related fruit, the common barberry (*Berberis vulgaris* L.), is a native of Europe but has been introduced to North America as an ornamental. It is a bushy, deciduous shrub with many arching, spiny branches and finely toothed oval leaves. The scarlet to orange-red berries, up to 1 cm in length, are similar to Oregon grapes in taste and can be substituted for them in any recipe.

## Where to Find
Oregon grapes are restricted in distribution to western North America. In Canada they occur only in British Columbia. *Berberis aquifolium* is widespread throughout the southern part of the province in open, dry, rocky areas, and *B. nervosa* is common in the southern coastal region in light to shaded woods. Common barberry is widely established in thickets and waste places, especially in southern Ontario and Quebec.

# Oregon Grapes
(Barberry Family)

*Berberis aquifolium* **Pursh** and *B. nervosa* **Pursh**
(Berberidaceae)

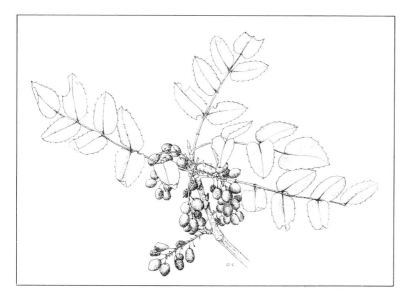

## How to Use
Oregon grapes ripen from midsummer to early fall, and are easy to harvest because the long, dense clusters can be broken off intact. However, the prickly leaves sometimes necessitate the use of gloves. The berries are too acidic to eat raw, but stewed with sugar or other fruits, or made into jams, jellies, and pies, they are very good. Like crabapples and cranberries they can be used to enhance the flavour of bland fruits, and can add a tanginess to any fruit dessert. They are a good source of vitamin C, can be made into a palatable wine, or can be prepared like lemonade to make a refreshing soft drink.

## Warning
Oregon grape and barberry plants, including the fruits, contain the alkaloid drug, berberine, used in medicine as an astringent for treating inflamation of the mucous membranes. This compound is potentially toxic if taken in large doses, and therefore it is recommended that you eat these fruits and foods made with them in moderation, and do not consume them as a regular part of your diet.

*Berberis aquifolium*

## Suggested Recipes

### Oregon Grape-ade

| 250 mL | ripe Oregon grapes | 1 cup |
|---|---|---|
| 1 L | water | 4 cups |
| 50 mL | sugar | ¼ cup |

Wash the grapes and place them in a blender with half the water. Blend until completely pulverized, then strain through a sieve or jelly cloth. Add the rest of the water and the sugar to the strained liquid and stir until the sugar is dissolved. Serve over ice cubes with a sprig of mint or slice of lemon as a garnish. Makes about 8 small glasses.

### Oregon Grape Pie

| | pastry for 2-crust pie | |
|---|---|---|
| 1 L | Oregon grapes, washed and drained | 4 cups |
| 250 mL | brown sugar | 1 cup |
| 5 mL | cinnamon | 1 tsp |
| 25 mL | flour | 1½ tbsp |
| 25 mL | butter | 1½ tbsp |
| 1 mL | salt | ¼ tsp |

Mix together all the ingredients for pie filling. Line pie pan with crust, prick with fork in several places, and pour in filling. Cover with top crust, prick with fork or knife, and bake at 200°C (400°F) for about 40 minutes, or until crust is crisp and golden brown.

## Oregon Grape–Apple Jelly

| | | |
|---|---|---|
| 1 kg | Oregon grapes | 2 lb |
| 125 mL | water | ½ cup |
| 500 mL | apple juice | 2 cups |
| 57 g | pectin crystals (1 package) | 2 oz |
| 2 L | sugar | 8 cups |

Wash Oregon grapes and place in saucepan. Add water and simmer, covered, for about 10 minutes, or until juice is free. Sieve the berries through a jelly bag. In a pot mix 750 mL (3 cups) of the strained juice with the apple juice and pectin. Stir well. Place over high heat and bring to a boil, stirring constantly. Add sugar and mix well. Continue stirring and bring to a full rolling boil. Boil hard for 2 minutes. Remove from heat, skim off foam, and pour into sterilized jelly glasses. Cover with hot paraffin and store in a cool place. Makes approximately 12 medium-sized jelly glasses. Good served with meat.

## Oregon Grape Wine

| | | |
|---|---|---|
| 1.5 kg | Oregon grapes | 3 lb |
| 500 g | raisins | 1 lb |
| 1.5 kg | sugar | 3 lb |
| 4 L | water | 1 gal |
| 8 g | wine yeast (1 package) | ¼ oz |
| | 1 slice dry toast | |

Do not use metal containers to make Oregon grape wine; the berries' acid may corrode metal. Pick fully ripe Oregon grapes, wash, and mash. Cut raisins in small pieces. Boil water and sugar together to make a syrup, then add Oregon grapes and raisins and boil 5 minutes longer. Place in a large open-mouthed container, cool to room temperature, then float yeast on the dry toast on the surface. Cover with cheesecloth. The next day, stir the mixture, and stir daily for 5 more days. Strain and place liquid in narrow-mouthed 4-litre (1-gal) bottle. Fit fermentation lock, and ferment for 3 weeks. Rack and ferment another 3 months, then bottle. Makes 4 L (1 gal). (Courtesy of Janet E. Renfroe.)

## More for Your Interest

The inner bark of Oregon grape stems and roots yields a bright-yellow dye that can be extracted by simply shredding the bark and boiling it in water. This dye was used by Indian peoples in British Columbia to colour basket materials, wood, feathers, and porcupine quills, and can also be used for cotton cloth and wool. Beautiful vivid hues can be obtained with this dye, but they tend to fade in sunlight or with repeated washing. The berries themselves can be used to make a reddish-purple stain, very difficult to remove from the hands and clothing.

The berries were eaten by most Indian peoples in British Columbia southwards, but seldom in any quantity. The Carrier Indians of central British Columbia used to eat the young and tender leaves, before the prickles became too sharp. They simmered them until soft in a small amount of water and ate them as a snack.

Common barberry is a second host of the stem rust, a serious fungus disease of grain crops such as wheat, oats, and barley. Efforts are being made to eradicate the plant, particularly in the Prairie Provinces, and it is now prohibited to import, propagate, or transport common barberry interprovincially.

# May-Apple

(Barberry Family)

## Other Names
Mandrake, wild jalap, raccoon-berry, wild lemon.

## How to Recognize
This perennial herb grows from a fleshy, spreading, underground rootstock to a height of 30 to 50 cm. The leaves are large and umbrella-like, with 5 to 9 prominent lobes spreading out like fingers on a hand. The leaf stem joins to the middle of the blade. The leaves are single on flowerless plants, paired on flowering plants. Borne between the leaves early in spring, the flowers are large and nodding, with 6 to 9 waxy, white petals. They have a rich, tropical fragrance, sometimes overpowering. The fruit is a fleshy, yellow, blotchy berry, oval-shaped and up to 5 cm long, with a sweet, somewhat musky odour. The outer skin, soft and pliant, encloses a mass of seeds in a gelatinous pulp. The fruits are at their prime in late summer, when the plants have begun to die down.

## Where to Find
May-apple grows, usually in large patches, in open deciduous woods, wet meadows, and along roadsides in southern Ontario and southern Quebec and throughout the eastern United States. It is sometimes cultivated as an ornamental and for its fruit.

## How to Use
Fully ripe may-apples have a distinctive flavour that some describe as delicately tropical—comparable to that of passion-fruit. The green parts of the may-apple plant, the rhizomes, seeds, and the unripe fruit are bitter and toxic (see Warning below), but the fully ripe fruits are safe, wholesome, and delicious. They can be eaten raw in moderate quantities (the seeds should be discarded), but are better when cooked and strained as a preserve. The juice can be extracted and made into a cooling drink: mix it with sugar in a light wine or squeeze it into lemonade or other fruit beverages.

---

*Podophyllum peltatum* **L.**
(Berberidaceae)

*Podophyllum peltatum*
(in leaf) (in fruit)

## Warning

The leaves, rhizomes, seeds, and unripe fruits of may-apple contain a resinous drug known as podophyllin, composed of a number of highly active and cathartic compounds. Ingestion can cause severe purging, gastroenteritis, and vomiting, and ingestion of the unripe fruit has been a frequent cause of poisoning in children. The ripe fruits, once the seeds are discarded, are the least toxic part of the plant and can be used in moderation. They have been eaten for centuries by native peoples and settlers and are especially loved by children, who must be careful to discard the seeds. Podophyllin has long been used in medicinal preparations for a variety of ailments. Animals do not normally consume this plant, but occasional browsing of the shoots has resulted in death to hogs, cattle, and sheep.

## Suggested Recipes

### May-Apple Marmalade

Cut the remnants of the blossoms from ripe, washed may-apples. Separate and set aside the skins from the pulpy seed mass. Cook the pulp in very little water for a few minutes until the seeds are loosened, then strain them out. Cut the skins into small pieces with scissors and add to strained pulp. Cook for 15 minutes, stirring frequently.

Measure fruit. Add 500 mL (2 cups) sugar and 15 mL (1 tbsp) lemon juice for each 250 mL (1 cup) fruit. Mix, bring to a boil, and cook to the consistency of thick syrup. Seal in sterilized jars. One litre of fruits will yield 8 medium-sized jelly glasses. (From Constance Conrader, "Wild Harvest".) Note: For may-apple jam additional pectin is necessary. Prepare the fruit as above, but strain the cooked fruit to make a purée. Then follow the recipe recommended for blueberry jam on any commercial pectin package.

## May-Apple Glacé Pie

| | | |
|---|---|---|
| 250 mL | flour | 1 cup |
| 30 mL | sugar | 2 tbsp |
| 125 mL | soft butter | ½ cup |
| | 1 egg yolk | |
| 5 mL | grated orange rind | 1 tsp |
| 500 gm | canned pears | 1 lb |
| 10 mL | plain gelatin | 2 tsp |
| 50 mL | orange juice | ¼ cup |
| 125 mL | rosé wine | ½ cup |
| 250 mL | may-apple marmalade (see previous recipe) | 1 cup |
| | whipped cream for topping | |

Make a cookie-dough pie crust with the first five ingredients, adding a small amount of water if necessary to bind the dough. Roll out pastry and fit into a medium-sized pie pan. Prick the bottom. Place in oven pre-heated to 180°C (350°F). After 5 minutes prick any bubbles that have formed under the crust and press the crust back into the pan. Continue baking for another 10 minutes, or until the crust is light golden. Cool.

Empty the pears into a strainer and let drain while you prepare the glaze. Soften gelatin in orange juice, add wine, and place on low heat until gelatin is melted. Remove from heat. Add may-apple marmalade, and mix well. Chill until gelatin begins to set. Slice pears and place in an even layer in the cooled pie shell. Pour chilled gelatin over the pears. Refrigerate until glaze is set. Cover with whipped cream just before serving.

This is a rich, elegant pie. You can prepare it a day ahead, for the flavours mellow, and the crust does not get soggy. (From Constance Conrader, "Wild Harvest".)

### More for Your Interest

Because of the strongly purgative effect of the drug podophyllin, found in may-apple, this plant was one of the principal medicinal remedies of the eastern North American Indians. It is still used as a drug to treat venereal warts.

# Hazelnuts

(Birch Family)

**Other Names**
Wild filbert, wild cobnut, hazel.

**How to Recognize**
These are both bushy deciduous shrubs up to 3 m high with greyish-brown bark and twigs that are fuzzy when young. The leaves are broadly oval and double toothed. The serrations of *C. americana* leaves are finer than those of *C. cornuta*. Male and female flowers are separate but on the same bush. The male, pollen-bearing flowers are borne in long, slender catkins that develop in fall and mature in early spring. The female flowers are small and inconspicuous, being surrounded by a cluster of reddish scales. The fruits, borne singly or in small clusters, ripen in autumn. Each consists of a smooth, oval, hard-shelled nut enclosed in a greenish, leafy sheath. The sheaths of *C. americana* nuts are soft to the touch and the ends are lobed and flared; those of *C. cornuta* are densely bristly towards the base and prolonged at the end into a long, slender beak, and for this reason it is sometimes called the beaked hazel. The nuts of both species closely resemble those of their relatives, the commerical filberts, but are smaller, thicker-shelled, and shorter.

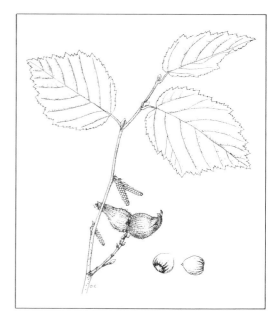

*Corylus americana* **Walt.**
**and** *C. cornuta* **Marsh.**
(Betulaceae)

*Corylus cornuta*

## Where to Find

Both species are found in moist woods and thickets. In Canada *C. americana* occurs only in the East, from southern Manitoba to southwestern Quebec, whereas *C. cornuta* ranges from British Columbia to Newfoundland.

## How to Use

Hazelnuts are difficult to harvest in quantity, especially in areas where the squirrel population gets them first. The job of removing the outer sheaths and shelling the nuts is a tedious one, but you will find the end product well worth your trouble. Indian people used to remove the sheaths by burying the nuts in damp mud for about twelve days. The sheaths would rot away, leaving the nuts ready to be cracked. Or you can spread the sheathed nuts in a cool, dry place to season and fully mature. To save space, the nuts can also be hung up in small quantities in cloth or paper bags. When dry, the sheaths can be pounded off with a hammer. The nuts can be cracked at the same time and the green kernels extracted. For most uses the kernels are best roasted in the oven at about 180°C (350°F) for about half an hour or until crisp and lightly browned. They can then be stored in sealed containers in a cool place until you are ready to use them. To use as table nuts, coat them lightly with cooking oil and sprinkle them with salt before roasting. The green (unroasted) nuts do not keep well after shelling and should be refrigerated or used immediately.

These wild nuts are every bit as tasty as filberts, and can be used in place of them in any recipe. They are excellent for fudge, chocolates, and other candies, and can be used in any muffin, cookie, or cake recipes calling for nuts. If you wish to follow Indian cooking traditions, try making hazelnut cakes or hazelnut soup.

## Suggested Recipes

### Hazelnut Cookies

| | | |
|---|---|---|
| 125 mL | butter | 1/2 cup |
| 125 mL | sugar | 1/2 cup |
| 250 mL | ground roasted hazelnuts | 1 cup |
| 250 mL | flour | 1 cup |
| | 1 egg yolk (save white for brushing tops) | |
| 2 mL | vanilla | 1/2 tsp |
| | 2–3 dozen whole hazelnuts | |

Cream butter until smooth, then gradually add sugar. Mix in remaining ingredients except the whole nuts, and knead until well mixed. Form into small balls and press a whole hazelnut on the top of each. Brush with the white of egg and bake on a greased cookie sheet at 160°C (325°F) for 10 to 15 minutes. Makes 2 to 3 dozen cookies, depending on size.

### Hazelnut Balls

| | | |
|---|---|---|
| | 1 egg white, stiffly beaten | |
| 500 mL | icing sugar | 2 cups |
| 15 mL | orange rind | 1 tbsp |
| | splash of brandy (optional) | |
| 500 mL | roasted hazelnuts, finely chopped | 2 cups |

Mix together egg white and icing sugar. Add orange rind and brandy (if desired) and continue to blend until creamy and smooth. Add nuts, mix well, and form into small balls. Dust with icing sugar, chill and serve. Makes about 3 dozen balls.

## Hazelnut Macaroons

| | | |
|---|---|---|
| | 2 egg whites, stiffly beaten | |
| 375 mL | icing sugar | 1½ cups |
| 125 mL | ground, roasted hazelnuts | ½ cup |

Fold the sugar and ground nuts carefully into the beaten egg whites. Form the mixture into small balls and place on waxed paper about 3 cm apart. Bake at 160°C (325°F) for 5 to 10 minutes, or until lightly browned. Watch carefully that they do not burn. Makes about 2 dozen cookies.

## Hazelnut Cakes, Indian Style

| | | |
|---|---|---|
| 500 mL | unroasted hazelnuts | 2 cups |
| 500 mL | water | 2 cups |
| 75 mL | cornmeal | ⅓ cup |
| 5 mL | salt | 1 tsp |
| 50 mL | vegetable oil | ¼ cup |

Grind nuts in a food grinder or blender, mix with water, and bring to a boil. Reduce heat and simmer for 30 minutes, stirring frequently. Mix in cornmeal and salt and let stand for 20 minutes, or until thick. Heat oil in a heavy skillet until it starts to sizzle. Drop the nut batter onto skillet with a large spoon. Allow cakes to cook and brown on one side, then flip over and brown on the other side. Serve hot or cold with butter and honey or maple syrup. Makes about 1 dozen small cakes.

## Hazelnut Soup

| | | |
|---|---|---|
| 500 mL | ground unroasted hazelnuts | 2 cups |
| 1 L | beef *or* chicken broth | 4 cups |
| | 1 medium onion, diced | |
| 25 mL | parsley, chopped | 1¹/₂ tbsp |
| | salt and pepper | |

Mix together all ingredients in large saucepan, bring to a boil, then simmer gently, stirring occasionally, for 1 hour. Serve hot. This is a rich soup; only small servings are necessary. Serves 4–6.

## More for Your Interest

Hazelnuts have been used for centuries by North American Indians, who used to search for the nut caches of squirrels to save themselves the trouble of harvesting and cleaning the nuts. Usually a handful of corn or other food would be left in place of the nuts so that the squirrels would not go hungry. Hazelnuts were a common Indian trade item. They were often pounded with berries, bulbs, or animal fat to make cakes, or were boiled in water to extract their oil, which was then used to flavour soups and dress vegetables. The straight, young shoots of hazel wood were used to make arrows, and the buds and roots were used in some areas to make a blue dye.

The European hazel, or filbert, (*Corylus avellana* L.) was grown by the early Greeks and Romans. Today the best filberts come from Spain and are known as Barcelona nuts. They are baked in large ovens to ensure their preservation and are exported to markets the world over. Cobnut is a name applied to one variety of filbert especially common in England. In Britain and Europe, hazels are planted in hedgerows, and the slender shoots are used for making hurdles, crates, and thatch-fastening rods, as well as baskets and wicker fences. Hazel charcoal is used by artists, and hazelnut oil is used as a base for fragrant oils and perfumes.

## Other Name
Elder.

## How to Recognize
There are at least three species of elderberry in Canada. Of these, two—the blue elder, or blue-berried elder, (*S. cerulea* Raf.) and the Canada elder, or common elder, (*S. canadensis* L.)—are the most important for our purposes, because of their readily edible fruit. The third species—red elder, or red-berried elder, [*S. racemosa* L. var. *pubens* (Michx.) Koehne]—has fruits that are very bitter and possibly poisonous, although they were a common food of the coastal Indians of British Columbia. A black-fruited variety of the red elder [*S. racemosa* var. *melanocarpa* (Gray) McMinn], sometimes treated as a separate species, is abundant in the Rocky Mountain region of British Columbia and Alberta. Some botanists prefer to designate blue elder as *S. glauca* Nutt. ex T. & G. Red elder is often separated completely from the typical European species, *S. racemosa*, and designated as *S. pubens* Michx.

# Elderberries

(Honeysuckle Family)

*Sambucus cerulea*

***Sambucus* species**
(Caprifoliaceae)

**53**

The elders are large, bushy, deciduous shrubs, sometimes, as in the case of blue elder, taking on the form of a small tree. The twigs are stout, with pale-green to greyish-brown bark. The woody part is very thin and surrounds a thick core of white pith. The leaves are large and pinnately compound, with usually 5 to 9 lance-shaped to elliptical pointed leaflets with finely but sharply toothed edges.

The flowers are small and creamy white, in dense clusters that are rounded or flat-topped in blue elder and Canada elder, and pyramidal or oval-shaped in red elder. The fruits are small, tightly clustered berries, ripening in late summer and fall in blue and Canada elders, and in early summer in red elder. The fruits of the blue elder have a distinct whitish waxy coating, or bloom, which gives them a greyish, smoky colour. Canada elder fruits are usually purple-black, although there are forms with red, green, or yellow fruits. Red elderberries are usually bright scarlet, but var. *melanocarpa* has blackish fruits, and forms exist with yellow, chestnut, or even white fruits.

## Where to Find
Elderberries grow in moist woods, clearings, and ravines, and along streams and roadways. In Canada, blue elder is found only in southern British Columbia from the east coast of Vancouver Island to the Rocky Mountain Trench area. Canada elder occurs in eastern Canada from Nova Scotia to Manitoba, and red elder grows across Canada from Newfoundland to British Columbia.

## How to Use
Fruits of the blue and Canada elders are edible when raw, although they have a somewhat rank taste and odour. Red elder fruits should never be eaten raw (see Warning below), but can be eaten without danger when cooked. Blue and Canada elder fruits are definitely improved by cooking. Somewhat bland and seedy when cooked alone, they are much better blended with other fruits, such as grapes, currants, lemons, and apples. They make good sauces, pies, muffins, and wine.

Elderberry juice can be extracted by gently simmering the berries in about one-quarter their volume of water for about 20 minutes. Mash them down, then strain through a jelly bag or several thicknesses of cheesecloth. This juice can be used as a beverage, mixed with other fruit juice and sweetened to taste, or to make syrup and jelly. Since elderberries are low in pectin, commercial pectin should be added if they are to be used alone to make jelly. However, when the juice is mixed half and half with the juice of crabapples or unripe wild grapes, or if a few cut-up apples are simmered with the berries, and a volume of sugar is added equal to the total volume of juice, an excellent, tangy jelly results without the addition of commercial pectin. Elderberries are exceedingly rich in vitamin C, containing more than either oranges or tomatoes.

Elder flowers also have many culinary applications. They are widely used in Europe and are becoming more popular in North America all the time. Some of their uses are outlined in the "More for Your Interest" section.

**Warning**

According to John M. Kingsbury, in *Poisonous Plants of the United States and Canada*, the green leaves, bark, and roots of elderberry plants contain substances that are purgative if taken in even moderate amounts. The roots may be responsible for death in hogs, and the young growth and leaves are believed harmful to cattle. For centuries children in Europe and North America have used the hollowed-out stems as blowguns and drinking straws, a practice that may cause sickness in some cases. The plants, including the fruit, contain many pharmacologically active substances: for example, various parts of the plant have been used medicinally to treat influenza and rheumatism. The ripe berries of blue and Canada elders are harmless, raw or cooked, but those of red elder, eaten raw, may produce nausea, and cases are known of people becoming sick after drinking red elderberry wine. Since unripe red elderberries are toxic, we recommend that you cook them before making wine. Use moderation with these berries as they have a strong laxative effect.

## Suggested Recipes

### Elderberry Wine

| | | | |
|---|---|---|---|
| 1.5 kg | ripe elderberries | 3 | lb |
| 500 g | raisins | 1 | lb |
| 5 L | boiling water | 4 | qt |
| 1 kg | sugar | 2 | lb |
| 8 g | dry yeast (1 package) | $1/4$ | oz |
| | 1 slice dry toast | | |

Put elderberries and raisins into a crock and pour the boiling water over. Press fruit well with a wooden spoon. Cover and keep for 4 days, stirring daily. Press berries through a fine sieve. Boil juice with sugar in saucepan for 30 minutes, stirring well. Pour back into crock and when liquid is lukewarm float the yeast, spread on a piece of toast, on top. Cover with cloth and leave for 3 weeks, or until fermentation (bubbling) has stopped. Bottle, corking lightly; after 2 weeks tighten the corks and leave for at least 3 months to mature before using.

### Blue Elderberry–Apple Pie

| | | | |
|---|---|---|---|
| 250 mL | blue elderberries | 1 | cup |
| 250 mL | apples, peeled, cored, and sliced | 1 | cup |
| 30 mL | cornstarch | 2 | tbsp |
| 50 mL | brown sugar | $1/4$ | cup |
| 5 mL | cinnamon | 1 | tsp |
| 2 mL | nutmeg | $1/2$ | tsp |
| | pastry for top crust on medium-sized baking dish | | |

Wash berries thoroughly and remove stems. Mix together with apples, cornstarch, brown sugar, cinnamon, and nutmeg, and spread out in a greased medium-sized baking dish. Top with pastry (or baking-powder biscuit dough), pricking at intervals with a fork. Bake at 180°C (350°F) for about half an hour, or until the top is golden brown. Serve hot with cream or milk.

## Elderberry Rob

| | | | |
|---|---|---|---|
| 1 L | elderberry juice | 4 cups | |
| 500 mL | sugar | 2 cups | |
| 15 mL | whole cloves | 1 tbsp | |
| 15 mL | allspice | 1 tbsp | |
| 15 mL | cinnamon | 1 tbsp | |
| 500 mL | brandy | 2 cups | |

Mix juice and sugar, add spices tied in cheesecloth, and boil for 20 minutes. Cool. Add brandy, pour into sterilized bottles, and cork. This beverage improves with age. You can also use mulberry, blackberry, or wild grape juice with this recipe. Without the brandy, these spiced syrups make pleasing bases for summertime beverages and ice-cream sodas. (From Constance Conrader, "Wild Harvest".)

## Autumn Medley Jam

| | | | |
|---|---|---|---|
| 1 kg | crabapples | 2 lb | |
| 1 kg | elderberries | 2 lb | |
| 1 kg | Italian plums | 2 lb | |
| 3 kg | sugar | 6 lb | |
| 15 mL | tartaric *or* citric acid | 1 tbsp | |

Peel, core, and slice apples, de-stem elderberries, and halve and de-stone plums. Place all fruits in a preserving kettle with just enough water to float them. Add tartaric *or* citric acid, bring to a boil, and simmer for 10 minutes, then add sugar. Stir until boiling and boil briskly until jam sets when tested (about 20 minutes). Pour into hot, sterilized glasses and seal with melted paraffin. Store in a cool place. Yields about 15 medium-sized jelly glasses.

## More for Your Interest

In Europe the dry flowering branches of elder are hung up in bunches in closets, basements, and lofts to repel rodents and insects such as fleas and cockroaches.

Elder-flower clusters are not only edible, but are delicious when cooked as fritters, or in pancakes or muffins. To make elder-flower fritters, simply dip the clusters, in full bloom, into an egg batter, drop into deep, hot oil and fry until delicately golden brown. Then dip into a sauce made of equal parts of honey and brandy, dust with cinnamon and icing sugar, and serve hot.

Wine can be made from elder flowers, used alone or mixed with various fruits. In European countries, fresh and dried flowers are used to make tea and a special syrup for coughs and sore throats. A good recipe for elder-flower wine is provided by Euell Gibbons in *Stalking the Wild Asparagus.* He also includes directions for making elder-flower face cream for improving the complexion. Elder-flower buds make an interesting pickle for cold meats and hot-dogs.

Elderberries were used by Indian peoples in many parts of Canada. They were often cooked into a sauce, then dried in cakes to eat during the winter. The Okanagan Indians of British Columbia used to spread out clusters of blue elderberries under a ponderosa pine tree just before the first snowfall. The berries would keep all winter under the snow, which took on a pink colouring over them, and could be dug up and eaten whenever required.

## Other Names

*Viburnum* species are variously called squashberry, hobblebush, moosewood, nannyberry, sheepberry, wild raisin, and withe-rod.

## How to Recognize

There are at least eight species of *Viburnum* in Canada. All are deciduous shrubs or small trees with relatively large, simple or lobed, opposite leaves, 10 to 15 cm long; small, white flowers (sometimes with a few large sterile flowers as well) in round-topped clusters; and fleshy "berries" (botanically, drupes), each with a single large, usually flattened, seed. The fruits range in colour from red to bluish black and all are edible, although some are better flavoured than others. Five species are mentioned here.

The high-bush cranberry (*V. opulus* L. var. *americanum*, also known as *V. trilobum* Marsh.) grows to a height of approximately 4 m, has distinctive leaves that are sharply and deeply 3-lobed, and bears large, red, acid fruits in drooping terminal clusters. The variety name distinguishes it from the less edible European variety known as the guelder rose, *gelderse roos*, or wayfaring tree, cultivated and sometimes escaped. The fruit of this variety is bitter and unpleasant. Another common species, squashberry [*V. edule* (Michx.) Raf., also known as *V. pauciflorum* La Pylaie], is a straggling shrub that grows up to 2.5 m high, with simple or, more often, shallowly-lobed leaves and large, red, acid fruits borne in small clusters at the leaf nodes. Hobblebush, or moosewood, (*V. alnifolium* Marsh.) has broadly oval-shaped leaves, finely toothed at the margins, and rust-coloured fuzz underneath on the leaf veins and on the young stems and leafstalks. It grows to a height of about 3 m. The fruits are red, turning darker with age, palatable and sweet, but with large seeds. Nannyberry, or sheepberry, (*V. lentago* L.) is a tall shrub, up to 10 m high, with simple, oval-shaped leaves, finely toothed at the margins, and with large, bluish-black fruits. The wild raisin, or withe-rod, (*V. cassinoides* L.) is very similar to nannyberry except that it has long-stalked flower and fruit clusters (those of nannyberry are sessile), and grows to only about 4 m in height.

# High-Bush Cranberry and Relatives

(Honeysuckle Family)

*Viburnum opulus* L. var. *americanum* Ait. and other *Viburnum* species
(Caprifoliaceae)

**59**

## Where to Find

In general the viburnums prefer moist woods or sometimes swampy areas. High-bush cranberry (*V. opulus* var. *americanum*) is found throughout the southern part of Canada from Newfoundland to central British Columbia. Squashberry is more northern in distribution, extending from Labrador to British Columbia and Alaska, commonly in boreal and montane forests. Hobblebush grows from Nova Scotia to the Great Lakes, extending southwards. Nannyberry occurs from Quebec to Manitoba, and wild raisin from Newfoundland to Manitoba.

## How to Use

The fruits of high-bush cranberry and squashberry are juicy, but quite acid. When first mature, they are hard, crisp, and sour, but after being subjected to a heavy frost they become soft and quite palatable, though still tart. They are best when cooked as a sauce or jelly, and indeed, when properly prepared, are equal in flavour to the true wild cranberry (*Vaccinium macrocarpon* and *V. oxycoccus*). High-bush cranberry sauce is excellent with meat and game. The ripe fruits of hobblebush, nannyberry, and wild raisin are quite sweet and flavourful even in a raw state, having been compared by some to raisins or dates in taste. Unfortunately, their large seeds make them difficult to use in cooking.

## Suggested Recipes

### High-Bush Cranberry Sauce

| | | |
|---|---|---|
| 1 L | high-bush cranberries, washed and de-stemmed | 4 cups |
| 50 mL | water | 1/4 cup |
| 15 mL | gelatin | 1 tbsp |
| 250 mL | sugar | 1 cup |

Place berries and water in a large saucepan, heat, and simmer until juice is free (about 10 minutes). Press through a sieve to remove skins and seeds. While still hot add gelatin and sugar and stir until dissolved, then cool. Serve with wild game or lamb. This sauce can be stored in the refrigerator and used as needed. Makes about 750 mL (3 cups) sauce.

### High-Bush Cranberry Jelly

| | | |
|---|---|---|
| 2 L | high-bush cranberries, washed and de-stemmed | 8 cups |
| 250 mL | water | 1 cup |
| | sugar | |

Place berries and water in a large saucepan, heat, and simmer until juice is free (about 10 minutes). Place in a fine nylon jelly-bag and allow juice to drain for several hours, or until dripping ceases. Measure juice and boil in saucepan, uncovered, for about 5 minutes. Measure sugar, allowing 1.5 L for every litre of juice (6 cups of sugar for every 4 cups juice), add to juice, stir until dissolved, then bring to a boil and cook, stirring constantly, until jelly sets when a small amount is tested on a cold plate. Pour into hot, sterilized glasses, cool, seal, and label. Store in a cool place. Makes approximately 6 medium-sized jars of a tangy, dark-red jelly.

## More for Your Interest

The viburnums were eaten by native peoples throughout Canada, although nowhere were they more important than on the British Columbia coast, where squashberries (*V. edule*) were greatly valued and used in large quantities. Squashberry patches in some areas were owned by certain families and passed on from generation to generation. The berries were preserved in oil or water in tall cedarwood boxes, and were commonly eaten at large feasts. The boxes of berries were also used as gifts or as trade goods. Among the Kwakiutl Indians of northern Vancouver Island, a box of squashberries was considered equal in value to two pairs of blankets.

The bark of the high-bush cranberry and squashberry was sometimes used in native smoking mixtures. The Norwegians and Swedes eat high-bush cranberries cooked with flour and honey, and distill a spirit from them. They were a favourite dish of Maine lumbermen, who used to eat them with molasses.

# Canadian Bunchberry

(Dogwood Family)

## Other Names
Bunchberry, dwarf dogwood, pigeonberry, dwarf cornel.

## How to Recognize
When blooming, this small perennial herb looks startlingly like a miniature flowering dogwood. The stems are wiry and erect, usually only 10 to 20 cm high, each bearing a cluster or whorl of leaves at the top, with 1 or 2 pairs of smaller leaves below. The smooth-edged leaves are elliptical, with prominent veins. The flower heads are solitary, each with 4 conspicuous white or creamy bracts (floral leaves) subtending the yellowish-green flower cluster. The fruits are borne in a tight cluster. Botanically termed *drupes*, they are the size of small peas, bright red and fleshy, each with a single, hard seed in the middle.

A closely related species, *C. suecica* L., resembles *C. canadensis* in general habit but has 3 or more pairs of relatively large stem

*Cornus canadensis* **L.**
(Cornaceae)

leaves, and those at the top are not as strongly whorled. There are several other minor differences, but as the fruits of both are similar and edible we will not distinguish further. The two species sometimes hybridize where their ranges overlap.

## Where to Find
The Canadian bunchberry is found in moist coniferous woods and bogs throughout Canada. It often grows in thick patches on stumps and rotting logs. *C. suecica* is found in similar habitats on the Pacific and Atlantic coasts and is also common in northern Europe.

## How to Use
Very few people realize that bunchberry fruits are edible. The pulp is quite juicy and pleasant tasting, although the hard seed at the centre makes the fruits somewhat difficult to eat. Bunchberries have the advantage of being easy to collect in quantity where they are abundant because the entire fruit clusters can be pulled off at once. The fruits ripen in summer and often stay on the plants through the autumn. When the seeds are removed or strained out, the pulp can be mixed with other fruits or used alone in cooking. Bunchberries are said to be an excellent ingredient for steamed plum puddings.

## Suggested Recipes

### Bunchberry Sauce

| 1 L | bunchberries | 4 cups |
|---|---|---|
| 125 mL | water | ½ cup |
| 50 mL | sugar | ¼ cup |
| 2 mL | salt | ½ tsp |
| 5 mL | cinnamon | 1 tsp |
| 15 mL | lemon juice | 1 tbsp |
| | sprinkling of grated nutmeg | |

Wash fruits and place with water in a saucepan over high heat. When water boils reduce heat and allow to simmer about 10 minutes, or until the fruits are very soft. Remove from heat and press through a sieve, or use a food strainer to remove the seeds and skins. Return the pulp to the saucepan and add the other ingredients, stirring gently over low heat until sugar is dissolved. Serve with pork, or as a dessert with ice cream. This recipe yields about 500 mL (2 cups).

## Northwest-Coast Indian Bunchberries with Eulachon Grease

| 2 L | bunchberries | 8 cups |
|---|---|---|
| 250 mL | water (first amount) | 1 cup |
| 500 mL | eulachon grease *or* melted lard | 2 cups |
| 250 mL | cold water (second amount) | 1 cup |

Place bunchberries and water (first amount) in large kettle, bring to a boil, and simmer about 5 minutes. Drain off water and cool fruits. In a large bowl mix eulachon grease (rendered from a small west-coast fish), *or* lard, with water (second amount) until thick and whitish. Mix in bunchberries and place in hot, sterilized jars (or containers lined with parboiled western skunk-cabbage leaves). Store in a cool place. To serve, scoop out mixture and sprinkle sugar over. The eater should discard the seeds himself. Makes about 1.5 L (6 cups).

## More for Your Interest

The Indians of the Pacific Coast used to eat bunchberries in large quantities and compared them to salal berries in taste. Among the Nootkan-speaking peoples of the west coast of Vancouver Island there is a legend that bunchberries originated from the blood of a young woman who was marooned at the top of a red cedar tree by her jealous husband. Wherever a drop of her blood touched the ground, a bunchberry plant grew. (Bunchberries are usually found around the base of cedar trees.)

Bunchberry is closely related to the flowering dogwoods (*Cornus florida* L. and *C. nuttallii* Audubon), both full-sized trees, and to a number of shrubby species, including the red-barked red-osier dogwood (*C. stolonifera* Michx.), whose range extends across Canada. The bright-red fruits of the flowering dogwoods look inviting, but have hard stones and thin, mealy, bitter-tasting flesh. Red-osier dogwood berries are white to light blue, and although bitter are quite juicy and, at least in British Columbia, were commonly eaten by Indians, usually mixed with other fruits such as saskatoon berries. Sometimes the hard stones were cracked open and the "nuts" inside eaten like peanuts.

## Soapberry and Buffaloberry

(Oleaster Family)

### Other Names
*S. canadensis* is also called soopollalie or russet buffaloberry, and *S. argentea* is called silver buffaloberry.

### How to Recognize
These are upright deciduous shrubs with opposite, smooth-edged leaves. Male and female flowers are borne on separate plants. Soapberry grows from 1 to 4 m high, and has oval-shaped leaves up to 6 cm long, green above and covered with a conspicuous coppery scurf on the undersides. The young twigs are also conspicuously brownish scurfy. The flowers are small and yellowish green, appearing before the leaves. The fruits are bright red-orange (occasionally yellow) and juicy, but extremely bitter.

Buffaloberry, which grows from 2 to 6 m tall, has silvery leaves and young twigs, and its branches are often spine-tipped. The leaves are generally smaller than those of soapberry and more rounded at the tips. The flowers are brownish and the berries brilliant red, fleshy, and somewhat bitter.

A related species, silverberry (*Elaeagnus commutata* Bernh.), is also silvery in appearance, with pinkish bark, yellow musky-smelling flowers, and large, dry silvery berries that are edible but too dry to be of much culinary use.

### Where to Find
Soapberry grows in open wooded areas, often on limestone soil, from British Columbia to Newfoundland. Buffaloberry grows along watercourses from eastern British Columbia to Manitoba. Silverberry grows along waterways, usually on gravelly soil, from British Columbia to Quebec.

*Shepherdia canadensis* (L.) Nutt. and *S. argentea* (Pursh) Nutt.
(Elaeagnaceae)

**67**

## How to Use

Soapberries and buffaloberries, though juicy and inviting in appearance, are usually too bitter to eat raw. However, in British Columbia, soapberries are used to make one of the most famous of Indian confections, named from its appearance but not its taste as "Indian ice cream". On the prairies, settlers used buffaloberries to make a piquant jelly. And both fruits can be made into a lemonade-like drink, very pleasant on a hot summer's day.

The name *soapberry*, or in Chinook jargon (the early trade language of the west coast) *soopollalie*, refers to the property of the berries of foaming up when beaten with water. Small amounts of saponin in the fruits cause this foaming, but concentrations of this compound are not high enough to make the berries dangerous to eat as long as they are taken in moderation. "Indian ice cream" is still made today as a special treat in many Indian households (see recipe, p. 70).

The term *buffaloberry* is said to be derived from the practice of some Plains Indians of flavouring their buffalo meat with a sauce of the berries. Buffaloberries are said to taste sweeter after a frost, and can be eaten raw at this time.

Both types of berries are small and soft and would be difficult to pick in quantity by hand. However, Indian peoples have devised a simple and effective way of picking them by poking the laden branches, still attached to the shrub, into a large container and whacking them sharply with a stick until all the ripe berries have fallen off into the container. Alternatively, a sheet or mat can be laid out underneath the bushes for the berries to drop on.

Both soapberries and buffaloberries can be dried, frozen, or canned. If canning, use only a little water with the berries, and no sugar. Once reconstituted, dried soapberries will whip up as well as fresh ones.

## Warning

Because of their high saponin content, soapberry and buffaloberry should be used in only small quantities, as saponins have a strong haemolytic action and cause intense irritation of the digestive system resulting in vomiting, abdominal pain, and diarrhoea when taken in large doses. However, man can tolerate small amounts of saponins as they are not readily absorbed from the gastrointestinal tract. They are widely present in many plants and in foods we normally consume: beets 0.3%, soybean seeds 0.5%, alfalfa seeds 2–3%, green tea 0.04%. They are also used as food additives. There is little information available as to the type or quantity of saponins present in *Shepherdia* fruits, but from their foaming properties it is assumed that the concentration is substantial. From our own experience and from the long history of use by Indian peoples in western Canada, the berries are not harmful in limited doses.

## Suggested Recipes

### "Indian Ice Cream"

| 250 mL | fresh soapberries | 1 cup |
|--------|-------------------|-------|
| 50 mL  | water             | $\frac{1}{4}$ cup |
| 50 mL  | sugar             | $\frac{1}{4}$ cup |

Canned, frozen, or dried soapberries may be substituted for fresh ones, in which case fewer will be needed to make the same amount of foam. In a clean glass, porcelain, or metal bowl, combine soapberries and water. It is important that the soapberries do not come in contact with any type of oil or grease or they will not whip. (Don't even use a plastic container for picking, as plastic tends to become greasy easily.) Beat with an egg beater, or in the traditional way—with the hands or a bundle of salal, thimbleberry, or maple leaves—until a foam is formed. Gradually add sugar (in the early days different berries were used as a sweetener) and continue beating until the foam is stiff. Serve immediately in bowls. Serves 4–6. This dish can also be made using apple juice instead of water, and less sugar.

## Buffaloberry Jelly

| | | |
|---|---|---|
| 1 L | buffaloberries, picked before the first frost | 4 cups |
| 50 mL | water | ¹/₄ cup |
| | sugar | |

Place berries and water in a pan, bring to a boil, and simmer for about 10 minutes, stirring frequently. Strain through a jelly bag, measure juice, and return to pan with an equal volume of sugar. Bring to a boil and keep boiling, stirring constantly, until jelly sets when tested on a cold plate. Pour into hot, sterilized jars and seal with melted paraffin. Store in a cool place. This is a beautiful light-orange jelly that can be used in place of cranberries on meat or chicken. Yields 4 medium-sized glasses.

## Buffaloberry-ade

| | | |
|---|---|---|
| 250 mL | ripe buffaloberries | 1 cup |
| 250 mL | sugar | 1 cup |
| 750 mL | boiling water | 3 cups |

Place berries and sugar in a large jar and add the boiling water. Stir until sugar has dissolved, mashing the berries at the same time. Let stand until cooled, then chill in refrigerator. When serving, pour through a strainer. Makes 8 small glasses.

**More for Your Interest**

Soapberry and buffaloberry have good potential as horticultural plants as they are both very attractive, especially when in fruit.

Even today, soapberries are widely sold and traded by natives in western Canada. They are believed to be rich in iron, and in some areas are made into tonics for blood ailments and acne. In a Bella Coola myth, the origin of soapberries is described: Raven, trickster and hero of Northwest Coast mythology, was invited to a feast by a certain mountain in the British Columbia interior, which was a chief with human traits. Soapberries flourished on the slopes of the mountain, but he wanted to keep them only for his guests. Raven was inside the mountain with the other guests—all different kinds of animals and birds. The mountain had closed all the entrances so that none of his guests could escape with the soapberries, but Raven used his power to make one of the guests go outside, and as soon as the door opened, Raven seized some of the soapberry whip and flew away, scattering drops of it as he flew. Wherever the drops fell soapberries grew, and from then on the Bella Coola people could make "Indian ice cream" whenever they wished. The mountain chief was very angry but could do nothing. This myth was recorded by T. F. McIlwraith in his work *The Bella Coola Indians.*

## Other Names
Black crowberry, curlewberry, crakeberry, monox, heathberry.

## How to Recognize
This low, woody, evergreen plant with stems up to 40 cm long, creeping and much branched, has short needle-like leaves growing in bottle-brush fashion along the twigs. Non-fruiting plants can easily be mistaken for heathers or miniature coniferous trees. The flowers are tiny and reddish, borne towards the ends of the twigs. The globe-shaped fruits (drupes) are fleshy, with several hard nutlets in the centre, and are usually deep, shiny black (sometimes purplish or white). The pulp is purple and juicy, slightly acid, or rather tasteless.

## Where to Find
Crowberry is widely distributed in Canada's subarctic region, from Newfoundland to the Yukon and northern British Columbia. It extends southwards in the mountains to the northeastern United States and in the West to northern California, and is also common in Alaska. Its distribution is circumboreal; it is widespread in northern Europe and the U.S.S.R. as well. It grows on exposed rocky bluffs and in peat bogs and muskegs, where it often covers the ground in dense mats.

# Crowberry
(Crowberry Family)

*Empetrum nigrum* L.
(Empetraceae)

**73**

## How to Use

The fruits are somewhat seedy and insipid but very juicy. A. E. Porsild, in "Edible Plants of the Arctic", states that crowberry is easily the most important fruit of the Arctic and, apart from cloudberry, the only one eaten regularly by the natives. Historically, crowberries have been little used by white people in North America although they are used by northern Europeans. They are a valuable emergency food source in the North, where there are few other berries available. They ripen in late summer and remain on the branches throughout the autumn. In some localities they are very plentiful and can be pulled off the stems a handful at a time.

They can be eaten raw as picked, but are best when mixed with other fruits, such as blueberries, and are said to improve with freezing. They can also be stewed with sugar and eaten with cream or ice cream. The Inuit and northern Indians usually dried or froze them for winter use. The frozen berries would be thawed and mashed together with seal blubber or whale oil to produce a kind of pudding, considered a great treat. When fermented, the juice of the crowberry is said to produce a good white wine.

## Suggested Recipes

### Eskimo Crowberry Pudding

| | | |
|---|---|---|
| 250 mL | young wild dock leaves *or* spinach | 1 cup |
| 50 mL | bacon drippings (*or* seal blubber if available) | ¼ cup |
| 250 mL | crowberries, fresh *or* frozen | 1 cup |

Cook the dock leaves *or* spinach until soft in a minimum of water, and add the bacon grease *or* seal blubber. (If you use seal blubber, warm it first.) Remove from heat, allow to cool, and mash until smooth and thoroughly mixed together. This mixture may be frozen if the crowberries are not available at the time it is made. When the crowberries are ripe add them to the greens (slightly thawed) and mix thoroughly with a spoon or knead with the hands. Freeze and use when needed as a snack or dessert. Mix with cooked de-boned cod-fish to make a complete, nutritious meal.

## Crowberry Cocktail

| | | |
|---|---|---|
| 1 L | crowberries | 4 cups |
| 500 mL | water | 2 cups |
| | juice of 1 lemon | |
| 75 mL | sugar | 1/3 cup |

Bring the crowberries and water to a boil. Reduce heat and allow to simmer for a few minutes, then mash to a soft pulp. Remove from heat and strain through a fine-meshed sieve or jelly bag to eliminate the seeds and skins. Cool. Add sugar and lemon juice, chill, and serve with a garnish on a hot summer day. Serves 4–6.

## More for Your Interest

The Scottish Highlanders use the crowberry sparingly. In Norway and Lapland crowberry wine is made, and in Siberia crowberry juice is drunk mixed with water. The people of the Kamchatka Peninsula eat the berries with fish, and make puddings by mixing them with the cooked bulbs of mission-bells (*Fritillaria camschatcensis*), a wild lily. The twigs were brewed as a tea by some Inuit groups, and the Haida Indians of the Queen Charlotte Islands boiled the branches with other wild herbs to make a medicine for tuberculosis, colds, and many other ailments. Explorer Samuel Hearne noted in his journal (p. 411) in the late eighteenth century that crowberries, or heathberries, ". . . are in some years so plentiful in Churchill, that it is impossible to walk in many places without treading on thousands and millions of them. . . . The juice of this berry makes an exceedingly pleasant beverage, and the fruit itself would be more pleasing were it not for the number of small seeds it contains." He further remarked that the Indians call them "grey-goose berries" because these birds relish them.

# Alpine Bearberries and Kinnikinnick

(Heather Family)

## Other Names
Alpine bearberries are called poisonberries in Newfoundland, where, despite the name, they are commonly eaten. Kinnikinnick is sometimes called mealberry, sandberry, hog-cranberry, or common bearberry.

## How to Recognize
There are two alpine bearberry species: black alpine bearberry [*A. alpina* (L.) Spreng.], and red alpine bearberry [*A. rubra* (Rehd. & Wilson) Fern.], sometimes considered a variety of the first. Both are low (usually under 10 cm high), densely branched dwarf shrubs with finely toothed leaves that are egg-shaped, tapering at the base and widest near the tip. Those of black alpine bearberry are thick, conspicuously net-veined, and 2 to 5 cm long; leaves of red alpine bearberry are thinner and usually slightly larger. The flowers of both species are white and urn-shaped, and the fruits are globular, pea-sized, and very juicy. The fruits are dark coloured in *A. alpina*, scarlet in *A. rubra*. The leaves of both species turn bright red in the fall, and in some regions of the Arctic the plants are so abundant they colour whole mountainsides.

***Arctostaphylos* species**
(Ericaceae)

*Arctostaphylos rubra*

Kinnikinnick [*A. uva-ursi* (L.) Spreng.] is also a low-growing shrub. It has long, flexible, trailing branches and shiny, leathery, evergreen leaves. The leaves are similar in shape to those of the alpine bearberry, but are somewhat smaller and are smooth-edged. The flowers are white to pinkish, usually in small clusters, and the red fruits are dry and mealy, with several hard nutlets fused together as a single stone.

**Where to Find**

Black alpine bearberry grows in the heaths and dry, open places of the mountains and on tundra from the Northwest Territories to Newfoundland. Red alpine bearberry is found in similar habitats in the Rocky Mountains and over much of northern Canada east to Newfoundland. Kinnikinnick is common in dry, sandy places and on rocky slopes throughout Canada except in the Far North.

## How to Use

When thoroughly ripe, alpine bearberries are juicy and pleasantly acid, though slightly bitter. They can be eaten raw but are better when cooked with sugar and eaten as a sauce. Most people prefer the red bearberries to the black ones. These are among the few berries to be found throughout much of northern Canada, and are thus valuable as a survival food. They can be gathered frozen during the winter and well into spring by clearing away the snow cover. Alpine bearberries are high in vitamin C and contain some vitamin A as well.

Kinnikinnick berries are not very good when eaten raw, but they are quite nourishing, being high in carbohydrates. They are improved considerably by frying in butter, oil, or fat to reduce the dryness. Like the alpine bearberries they are an excellent survival food because they remain on the plants through the winter and can be gathered by digging under the snow. They can also be dried and added to soups and stews.

## Suggested Recipes

### Bearberry Sauce

| | | |
|---|---|---|
| 250 mL | black *or* red bearberries | 1 cup |
| 125 mL | water | ½ cup |
| 125 mL | honey | ½ cup |

Place berries and water in saucepan, bring to a boil, and simmer 10 minutes. Press through a sieve, return to pan, stir in honey, and simmer 5 minutes more. Cool and serve on ice cream or pancakes or with meat.

## Kinnikinnick Berries with Dumplings

| | | |
|---|---|---|
| 250 mL | flour | 1 cup |
| 5 mL | baking powder | 1 tsp |
| 1 mL | salt | ¼ tsp |
| 30 mL | mountain goat fat *or* lard | 2 tbsp |
| 75 mL | milk *or* water | ⅓ cup |
| 500 mL | kinnikinnick berries | 2 cups |
| 250 mL | boiling water | 1 cup |
| 30 mL | sugar | 2 tbsp |

Mix together flour, baking powder, and salt, and work in fat *or* lard with the fingers. Add enough milk *or* water to form a stiff dough, knead lightly, and roll into small balls with the hands. Add kinnikinnick berries (fresh, or dried and reconstituted) and sugar to the boiling water, and drop the "dumplings" in one at a time over the berries. Cover and let steam about 20 minutes, or until dumplings are no longer sticky inside. Serve hot. Serves 4. (A similar dish was made by the Bella Coola Indians of British Columbia.)

## More for Your Interest

Kinnikinnick leaves are among the most famous of all wild smoking substances. They are still used today by some Indians and trappers, alone or mixed with real tobacco. The leaves can also be made into a pleasant tea, as described in our *Wild Coffee and Tea Substitutes of Canada*, but this should be used in moderation, because it may cause liver disease with frequent use, and it can cause gastrointestinal upsets in children because of the high tannic acid content. Kinnikinnick leaves can be used to produce a soft yellow-grey dye, requiring the initial use of an alum mordant.

Kinnikinnick is an excellent ground cover since it is evergreen, has brightly coloured berries, propagates easily, and is tolerant of a variety of habitats. It is being employed more and more in landscaping in Canada.

# Salal

(Heather Family)

## How to Recognize

Salal is an erect to partially creeping, freely branching, evergreen shrub up to 2.5 m tall. The stems are flexible and tough, and are able to withstand heavy, lasting snows, springing back into an erect position when the snow melts. The shiny, leathery leaves are oval-shaped and pointed at the tips, with finely saw-toothed edges. The flowers, pinkish to white, and urn-shaped, are borne in long, one-sided clusters. The blooming season ranges from early spring to mid-summer. The fruits are nearly black, hairy, thick-skinned, and berry-like, with numerous tiny seeds. At the tip of each fruit is a conspicuous star-shaped depression. The fruits remain on the branches for several weeks in good condition, giving the forager a long season to enjoy them.

Several other species of *Gaultheria* occur in Canada, including the well-known wintergreen (*G. procumbens* L.) and the delicious creeping snowberry, or capillaire, [*G. hispidula* (L.) Muhl.]. The latter is sometimes placed in a separate genus, *Chiogenes.* The fruits of these two species have a mild wintergreen flavour. Wintergreen berries are bright red and globular; those of creeping snowberry are white and oblong. (**Warning:** Oil of wintergreen, the compound that gives wintergreen leaves and berries their aromatic fragrance and taste and, to a lesser extent, flavours the leaves

*Gaultheria shallon* **Pursh**
(Ericaceae)

and berries of creeping snowberry, is chemically known as methyl salicylate, which is closely related to aspirin. It is known to be toxic in high doses and should be strictly avoided by anyone who is allergic to aspirin. However, the berries of these plants and tea made from the leaves should not be harmful when used in moderation by those with a normal tolerance for them. Children should avoid eating the berries.)

## Where to Find
Salal is found only in British Columbia, in coastal forests up to 55°N latitude, and from sea level to 800 m elevation, and again sporadically in the Kootenay Lake area. It commonly grows on rotten logs and stumps, and often forms dense, impenetrable thickets in moist, shady habitats along the coast. Wintergreen grows in woods and clearings from Newfoundland to Manitoba. Creeping snowberry is found in peat bogs and muskegs and wet coniferous woods from Newfoundland to central British Columbia.

## How to Use
Salal is one of the best of the western wild fruits, being sweet, juicy, and flavourful. As with many fruits the quality varies with locality and weather conditions, so do not be discouraged if you sampled the berries once and did not care for them; try again in another location. The berries are often very abundant and can be picked quickly and easily by breaking off the long clusters with stems still attached. Some people do not like the tiny seeds, but in our opinion succulent salal berries that have ripened in warm, sunny weather are hard to beat. They can be eaten raw or used as an ingredient in jellies, jams, syrups, and pancakes. They can also be used fresh or frozen to make pies, they mix well with other fruits, and they can be dried and used like raisins in cookies and fruit cakes. Their great abundance makes them an excellent survival food in season.

Working in a remote area of Vancouver Island during a very hot summer, Adam Szczawinski once experienced firsthand the value of salal berries as an emergency food source. He had been camping only one week when an old black bear paid a visit and helped itself to all his fresh fruits and vegetables. Fortunately the area was thickly covered with salal, and Operation Salal was quickly implemented. The berries were gathered in quantity and made into every form of edible product imaginable, with the help of some sugar (which, having been kept in a metal box, survived the bear's visit). Of all the different concoctions he made, there is one he will never forget—a refreshing and satisfying salal drink mixed up each morning from a stock supply of salal syrup, stored all day at the bottom of a cool, clear stream. This drink was so delicious that he continued to use it even after extra food supplies had been delivered.

## Suggested Recipes

### Vancouver Island Salal Syrup

| 1 L | water | 4 cups |
|---|---|---|
| 1 kg | sugar | 2 lb |
| 2 L | salal berries | 8 cups |

Combine water and sugar in a saucepan, stir to dissolve sugar, and heat to boiling. Boil for 5 minutes, then add berries and cook until the skins pop (about 5 to 7 minutes). Remove from heat, strain, and chill.

To make a drink from this syrup, simply dilute with water (try 1 part concentrate to 3 parts water). Pour into jar and store at the bottom of a creek, or in the refrigerator, until icy cold.

### Salal–Rhubarb Jam

| 1 kg | rhubarb | 2 lb |
|---|---|---|
| 250 mL | water | 1 cup |
| 1 kg | salal berries | 2 lb |
| 2 kg | sugar | 4 lb |

Wash rhubarb and cut into short segments. Simmer gently with the water until soft and pulpy. Add salal berries and sugar. Bring to a boil and boil briskly for 10 minutes, stirring frequently. Remove from heat, and pour into hot, sterilized jars. Seal with melted paraffin and store in a cool, dark place. Makes about 18 medium-sized glasses.

## Salal–Black Currant Pie

|  | pastry for 2-crust pie |  |
|---|---|---|
| 250 mL | sugar | 1 cup |
| 50 mL | flour | 1/4 cup |
|  | dash of salt |  |
| 500 mL | salal berries | 2 cups |
| 250 mL | black currants | 1 cup |

Line pie plate with pastry and prick with a fork. Combine sugar, salt, and flour, and place half of the mixture in the pastry shell. Combine the fruit, place in the shell, and sprinkle with the balance of the flour and sugar mixture. Cover with top crust, seal edges, and prick the top with a fork. Bake at 200°C (400°F) for 30 minutes. If black currant jam is substituted for black currants, omit the sugar,

## Cherry-flavoured Salal Pie

|  | pastry for 2-crust pie |  |
|---|---|---|
| 750 mL | salal berries | 3 cups |
| 250 mL | sugar | 1 cup |
| 50 mL | flour | 1/4 cup |
| 2 mL | vanilla extract | 1/2 tsp |
| 2 mL | almond extract | 1/2 tsp |
|  | pinch of salt |  |
| 125 mL | water | 1/2 cup |

Line pie plate with pastry and prick with a fork. Mix the salal berries, sugar, and flour together. Add flavourings, salt, and water. Mix well, pour into pastry shell, cover with top crust, seal edges, and prick the top with a fork. Bake at 200°C (400°F) for 30 to 35 minutes. The almond extract gives the pie the flavour of fresh cherries.

## More for Your Interest

Salal berries were a favourite of the Indians of the Northwest Coast. They gathered them in large quantities and ate them fresh, mashed them with oil, or cooked them to a jam-like consistency using red-hot rocks, and dried them into flat cakes that could be stored for many months. In winter, when needed, they could be rehydrated by soaking in water overnight. Salal berries were often used to sweeten other berries, such as soapberries and currants, and were often eaten with oil, fish eggs, or salmon.

Salal was originally collected in 1792 by Captain George Vancouver's botanist, Archibald Menzies. In 1825, botanist David Douglas (after whom Douglas-fir is named) recognized this shrub on the Columbia River and recorded in his journal: "Gaultheria shallon (salal) was the first plant I took in my hands. So pleased was I that I could scarcely see anything but it. . . . [It] would make a valuable addition to our gardens" (Morwood 1973, p. 56). Thanks to Douglas, salal did become a well-known ornamental in Europe and Great Britain. Today, the shiny, leathery, evergreen foliage of salal is in wide demand in the florist trade, particularly in Christmas bouquets and wreaths.

The berries of both wintergreen and creeping snowberry were popular among the early settlers of eastern Canada and the United States, and were sometimes gathered and sold in local markets. Wintergreen berries and leaves are the original source of oil of wintergreen (methyl salicylate), used in flavourings, and they make a pleasant tea, as described in our *Wild Coffee and Tea Substitutes of Canada* (but see Warning, p. 80).

## Other Names

Moss cranberry, wild cranberry; *V. macrocarpon* is known as large, or American, cranberry, *V. oxycoccus* as small cranberry.

## How to Recognize

The opinions of botanists vary as to how many species of bog cranberries should be recognized. There may be as many as four in Canada. Most botanists prefer to include them as a subgroup of the genus *Vaccinium* (see wild blueberries and bilberries, p. 94), but since their flower structure is quite different from that of other vacciniums, some botanists prefer to designate them as a separate genus, *Oxycoccus*. Since the different species of bog cranberry are quite similar in appearance, we will treat them together here.

They are all low, slender, trailing, woody shrubs with very thin, flexible branches and tiny, oval-shaped, smooth-edged leaves. The tiny but distinctive flowers, borne on thread-like stems, are pink, nodding, and 4-parted, with recurved petals and exserted stamens. The berries, which can be 1 cm or more in length, are elongated or globular, bright red, many-seeded, and juicy, but quite acid. Considering the size of the plants, the berries are amazingly large. Those of *V. macrocarpon* are especially big; this is the species that has yielded the many cultivated forms of cranberry. Do not confuse these species with the low-bush cranberry (*V. vitis-idaea*), described in the next section, or with the completely unrelated high-bush cranberry (*Viburnum*), discussed on page 59. (No risk is involved in confusing them, however, since the fruits of all of them are edible.)

## Where to Find

These cranberries require wet, acid conditions, and are found in muskegs and peat bogs throughout Canada. *V. macrocarpon* occurs from Newfoundland to Manitoba, and *V. oxycoccus* is transcontinental.

# Bog Cranberries

(Heather Family)

*Vaccinium macrocarpon* Ait.
and *V. oxycoccus* L.
(Ericaceae)

## How to Use

Like the low-bush and high-bush cranberries, these fruits ripen rather late in the year. By happy coincidence, they are usually ready for harvesting around the Canadian Thanksgiving holiday, just in time to make a superb sauce for the turkey. They are too tart to eat raw, although they become sweeter after being exposed to frost. They are quite high in vitamin C and very flavourful after being sweetened with sugar or honey.

Fresh cranberries store quite well under refrigeration if kept covered and unwashed without being bruised or pressed. To freeze, select deep-red, firm berries and pack dry without sugar. It is a good idea to quick-freeze them by spreading them out loosely on a tray in the freezer for 1 to 2 hours. When frozen hard, place in containers or freezer bags. When they are needed, rinse them quickly in cold water and use them in any recipe calling for fresh cranberries.

Cranberries can be dried in a slow oven with the door partly open, until they become quite dry and shell-like in appearance. To use the dried berries, soak them in water overnight and then use as you would fresh berries.

## Suggested Recipes

### Cranberry Pudding

| | | |
|---|---|---|
| 125 mL | brown sugar (first amount) | ½ cup |
| 250 mL | flour | 1 cup |
| 10 mL | baking powder | 2 tsp |
| | pinch of salt | |
| 250 mL | fresh cranberries | 1 cup |
| 125 mL | milk | ½ cup |
| 30 mL | vegetable oil | 2 tbsp |
| 250 mL | brown sugar (second amount) | 1 cup |
| 15 mL | butter | 1 tbsp |
| 5 mL | vanilla | 1 tsp |
| 500 mL | boiling water | 2 cups |

In a casserole mix the brown sugar (first amount), flour, baking powder, and salt. Add cranberries, milk, and oil, and mix in, spreading out evenly in the bottom of the casserole. On top of this batter put the brown sugar (second amount), butter, and vanilla. Pour boiling water over and bake 45 minutes at 180°C (350°F). Serves 4–6. (Courtesy of Marilyn Hirsekorn.)

## Cranberry Pie Filling

| | | | |
|---|---|---|---|
| 50 mL | minute tapioca | $^1/_4$ | cup |
| 250 mL | water | 1 | cup |
| 625 mL | white sugar | $2^1/_2$ | cups |
| 125 mL | raisins | $^1/_2$ | cup |
| 1 L | fresh cranberries | 4 | cups |
| | grated rind of 1 orange | | |

Boil tapioca in water until soft. Add sugar, raisins, berries, and orange rind, bring to a boil, and boil covered for about 3 minutes. Remove from heat and use as pie filling. Makes filling for 2 pies. (Courtesy of Marilyn Hirsekorn.)

## Cranberry Cocktail

| | | | |
|---|---|---|---|
| 500 mL | fresh cranberries | 2 | cups |
| 500 mL | water | 2 | cups |
| 75 mL | sugar | $^1/_3$ | cup |
| 500 mL | orange juice | 2 | cups |
| | small bottle of ginger ale *or* soda water | | |
| 125 mL | vodka *or* gin (optional) | $^1/_2$ | cup |

In a saucepan, cook cranberries with water until the berries pop (about 5 minutes). Strain through a sieve or cheesecloth. Boil juice for 2 to 3 minutes, then combine with sugar and orange juice. Chill. Add ginger ale *or* soda water, and vodka *or* gin if desired, just before serving. Makes 10 small servings.

## Thanksgiving Cranberry Sauce

| | | |
|---|---|---|
| 1 L | fresh cranberries | 4 cups |
| 30 mL | butter | 2 tbsp |
| 500 mL | sugar | 2 cups |
| 500 mL | water | 2 cups |
| | juice of 1 lemon | |
| 5 mL | cinnamon | 1 tsp |

Combine cranberries, butter, sugar, and
water in a saucepan. Heat to boiling point,
stirring until sugar dissolves. Boil rapidly
until berries pop (about 5 minutes). Add
lemon juice and cinnamon. Cover and
refrigerate until firm. Keep refrigerated or
freeze until needed. Serve with poultry.
Yields about 1 L (3 to 4 cups) of sauce.

## More for Your Interest

The alternate genus name for bog cranberries, *Oxycoccus*, is Latin for "sour berry".

According to Barbara R. Fried, in *The
Berry Cookbook*, these berries keep so well
that 10 barrels of them were shipped back to
England by the early Massachusetts
colonists as a gift for King Charles II. The
colonists originally called them "craneber-
ries" because of the resemblance of the
blossoms and stems to the head and neck of
a crane.

Commercial cranberries, derived from
forms of *V. macrocarpon*, are grown in habitats similar to that of their wild relatives—
flat, rich peat marshes and muskegs. The
land must be carefully controlled for mois-
ture with sufficient water in storage to flood
the fields quickly, since the plants are
sensitive to drought or frost. The plants are
propagated by cuttings. Before planting, the
soil is covered with a few centimetres of sand
to keep down the weeds and help the cuttings to root. After three years the plants
bear a full crop, and after five years a fresh
layer of sand is laid down for a new planting.
The growers often harvest the berries by
flooding the fields and allowing the ripe
fruits to float to the surface, where they can
be skimmed off.

# Low-Bush Cranberry

(Heather Family)

## Other Names

Lingonberry (or lingenberry), foxberry, partridgeberry, European cranberry, mountain cranberry, alpine cranberry, rock cranberry.

## How to Recognize

This plant is a low, creeping, dwarf shrub with small, leathery, lustrous, evergreen leaves that are oval-shaped, rounded at the tips, rolled under at the margins, and dotted beneath. The flowers are pinkish and urn-shaped, in compact terminal clusters. The berries are bright red, relatively small, and quite acid. They ripen in autumn and often remain on the plants over the winter.

## Where to Find

This little shrub grows in acid peat bogs and muskegs, often at considerable elevations in the mountains, and is especially common throughout our northern forests.

## How to Use

These cranberries are seldom eaten raw, as the fruit is too tart and acid. However, they represent one of the most important fruits of the North and are fairly rich in vitamin C. The taste may be improved by adding sugar or blending them with other fruits. Alone or mixed, the berries make good jellies, jams, sauces, pies, tarts, and cakes. The most common use is as a substitute in sauces for regular wild or commercial cranberries (see *Vaccinium macrocarpon* and *V. oxycoccus*, p. 85). They are well known and widely used in Europe, and are much more important there than the regular cranberries so popular on our continent.

A big advantage of the low-bush cranberry is that the berries survive the winter well on the plants and, harvested, may be kept in storage in a cool place for a very long time. Animals and birds can dig them out from under the snow and utilize them as food during the winter.

Adam Szczawinski recalls that at his boyhood home in Poland the berries were collected in October, before snow covered the ground, placed in oak barrels filled with cold water, covered with a thick layer of straw, and placed outside in the open or in an unheated place. They were used all winter in desserts and baked goods. At Christmas time they were, along with apples, one of the most important fruits, traditionally being used with other stewed fruits at the Christmas Eve dinner.

*Vaccinium vitis-idaea* L.
ssp. *minus* (Lodd.) Hult.
(Ericaceae)

## Suggested Recipes

### Low-Bush Cranberry–Walnut Sauce

| | | |
|---|---|---|
| 250 mL | sugar | 1 cup |
| 250 mL | water | 1 cup |
| 500 mL | cranberries | 2 cups |
| | 1 large apple, peeled and sliced | |
| 50 mL | walnuts, coarsely chopped | $\frac{1}{4}$ cup |
| | juice of 2 lemons and 2 oranges | |
| 2 mL | powdered cloves | $\frac{1}{2}$ tsp |
| 2 mL | cinnamon | $\frac{1}{2}$ tsp |

Combine sugar and water in saucepan. Bring to a boil. Add cranberries and apple and simmer until cranberry skins pop (about 5 minutes). Remove from heat and add other ingredients. Cool and serve with roast chicken or turkey. This sauce can also be used as a delicious glaze for ham. Makes 500 mL (2 cups) of sauce.

### Low-Bush Cranberry–Almond Frappé

| | | |
|---|---|---|
| 250 mL | milk | 1 cup |
| 125 mL | cranberries | $\frac{1}{2}$ cup |
| | 1 orange, peeled and cut in pieces | |
| | small square of orange rind | |
| 50 mL | honey | $\frac{1}{4}$ cup |
| | 3 to 4 drops almond extract | |
| | pinch of salt | |

Put all ingredients in a blender. Cover and start on low speed, then turn to high speed and run until the fruits are completely liquefied and smooth. Pour into a freezing tray and freeze to a mushy consistency. Serve partially frozen as a good, healthy dessert. Serves 4.

## Lingonberry–Red Currant Jam

| | | |
|---|---|---|
| 250 mL | red currants | 1 cup |
| 50 mL | water | ¼ cup |
| 375 mL | sugar | 1½ cups |
| 500 mL | low-bush cranberries | 2 cups |

Stew the currants with water for 30 minutes, then strain off the juice. Add sugar and cranberries to the juice, stir well, and boil for 5 minutes. Then reduce heat and simmer for 20 minutes. Leave overnight in the pot. The next day boil for 10 to 15 minutes, or until jam sets when tested. Pour into hot, sterilized jars and seal with melted paraffin. Store in a cool place. This is a pleasantly tart jam, quite rich in vitamin C. Yields about 3 to 4 medium-sized glasses.

### More for Your Interest

The Haida Indians of British Columbia called these berries "dog-salmon eggs", and regarded them as the winter form of the bog cranberry (*Vaccinium oxycoccus*).

Low-bush cranberries are extensively harvested in northern Europe, northern Asia, and to a lesser degree in North America, and used in the same way as bog cranberries. They are known under many local names and there is always a certain amount of confusion about the different kinds of "cranberries". It is interesting that many of the colloquial names pertain to animals. The name *cranberry* itself is derived from *craneberry*, and the berries of this species are called "foxberry" in Nova Scotia, "partridgeberry" in Newfoundland, and "cowberry" in England. Germans, Norwegians, and Danes call them "tyttenberry", and in Sweden and Finland they are called "lingonberry" or "kroessaberry". In the United States they are commonly known as "European cranberry", and in Canada the most commonly used names are "low-bush cranberry", "rock cranberry", and "mountain cranberry". The Latin specific name *vitis-idaea* translates to mean "grape of Mount Ida".

In Scandinavian countries, the berries are gathered in large quantities, placed in barrels filled with cold water, and sold on world markets as a substitute for cranberries. Hudson's Bay Company explorer Samuel Hearne noted a similar practice in Canada around 1770 (p. 411):

> When carefully gathered in the Fall, in dry weather, and as carefully packed in casks with moist sugar, they will keep for years, and are annually sent to England in considerable quantities as presents, where they are much esteemed. When the ships have remained in the [Hudson's] Bay so late that the [low-bush] Cranberries are ripe, some of the Captains have carried them home in water with great success.

# Wild Blueberries, Bilberries and Huckleberries

(Heather Family)

## Other Names
Whortleberries, blaeberries.

## How to Recognize
There are at least eighteen *Vaccinium* species to be found in Canada. The fruits vary in colour from red to blue to black, and all of them are edible. Members of this genus are variously called blueberries, whortleberries, bilberries, huckleberries, and cranberries. In many cases these names are used interchangeably, although the true blueberries are usually distinguished by having berries in terminal clusters, whereas the bilberries and whortleberries have berries borne singly in the leaf-axils. All are sometimes called huckleberries, but this name is also applied to the berries of a different, related genus, *Gaylussacia*, whose fruits are seedier than those of the vacciniums, but can be used similarly.

For the purposes of this book we have treated the cranberries (*V. vitis-idaea*, p. 90), and (*V. macrocarpon* and *oxycoccus*, p. 85) separately, as their fruit is tarter and ripens later than most of the other vacciniums.

The blueberries and their relatives have bushes ranging in height from very low to 2 m or more tall, and their leaves and berries differ in size and shape, but all are recognizable to most people as belonging to the same group of fruits, and all can be used more or less interchangeably in cooking. A number of cultivated varieties have been developed from the wild blueberries and are widely marketed in North America.

Among the best-known species of wild blueberries, bilberries, and huckleberries in Canada are: Canada, or sour-top, blueberry (*V. myrtilloides* Michx.), low, with velvety, smooth-edged leaves and clustered, light-blue, tart berries; low-bush blueberry (*V. angustifolium* Ait.), very similar but with smooth, finely toothed leaves; bog bilberry (*V. uliginosum* L.) and dwarf bilberry (*V. caespitosum* Michx.), both low shrubs with singly borne berries, the first with smooth-edged leaves, the second with finely toothed leaves; high-bush blueberry (*V. corymbosum* L.), tall, with clustered blue berries; oval-leaved bilberry (*V. ovalifolium* Smith), medium-sized, with smooth-edged, oval-shaped leaves and sweet, light-blue, singly borne berries; mountain bilberry (*V. membranaceum* Dougl. ex Hook.), low to medium-sized, with elliptical, finely toothed

*Vaccinium* **species**
(Ericaceae)

*Vaccinium membranaceum*
*Vaccinium myrtilloides*

leaves and blackish, shiny, single berries; evergreen huckleberry (*V. ovatum* Pursh), medium-sized, with thick, evergreen leaves and tart, blackish to blue, late-ripening, clustered berries; and red huckleberry (*V. parvifolium* Smith), medium-sized with thin, deciduous leaves and pink to dark-reddish single berries. Some of these species have juicier, more flavourful berries than others, but all lend themselves to a wide variety of culinary uses.

## Where to Find

Blueberries, bilberries, and huckleberries occur throughout Canada, even in the Far North. Some, such as Canada blueberry and bog bilberry, are found in acid peat bogs and muskegs. Others grow in woods ranging from damp to dry, often at considerable elevations in the mountains. Many seem to grow well in burned or logged-over areas.

Canada blueberry is found from Labrador to British Columbia, low-bush blueberry from Labrador and Newfoundland to Manitoba, and bog bilberry across northern Canada, extending southwards in some areas. Dwarf bilberry occurs across Canada from Labrador to Alaska, but mainly in the South, and high-bush blueberry, which is widely cultivated, grows wild only in the southeastern part of Canada. Oval-leaved bilberry is common in British Columbia and occurs sporadically in the Great Lakes region and eastwards. Mountain bilberry is common in the mountainous areas of British Columbia and Alberta and recurs in the Great Lakes area. Evergreen huckleberry and red huckleberry grow in wooded areas of coastal British Columbia, where they are very abundant locally.

## How to Use

Blueberries, bilberries, and huckleberries are probably the most popular of all wild fruits in Canada. They are delicious in almost any form—eaten raw, baked in cakes, muffins, and pancakes, made into pies, jams, jellies, and syrups, and mixed with other fruits of all kinds. They freeze well, can be canned or dried and can also be made into wine. They can be substituted in any recipes that call for commercial blueberries, although you may find they require a little more sweetening.

Wild blueberries and their relatives are so widespread and plentiful in Canada that they are a valuable survival food.

The following recipes can be used for any of the blueberries, bilberries, or huckleberries that have been mentioned.

## Suggested Recipes

### Wild Blueberry Jam

| | | |
|---|---|---|
| 1 L | blueberries, mashed | 4 cups |
| 30 mL | lemon juice | 2 tbsp |
| 57 g | pectin crystals (1 package) | 2 oz |
| 1.25 L | sugar | 5 cups |

Put berries into a large saucepan. Mix in lemon juice and pectin crystals and place over high heat, stirring constantly, until mixture comes to a full boil. Stir in sugar, bring back to a full, rolling boil, then boil hard 1 minute, stirring constantly. Remove from heat, skim off foam, stir about 5 minutes to cool slightly, and ladle into sterilized jars or jelly glasses. Seal at once with melted paraffin. Label and store in a cool place. Makes about 8 to 10 medium-sized jelly glasses.

### Blueberry Buckle

| | | |
|---|---|---|
| 125 mL | sugar (first amount) | $\frac{1}{2}$ cup |
| 500 mL | flour (first amount) | 2 cups |
| 12 mL | baking powder | $2\frac{1}{2}$ tsp |
| 1 mL | salt | $\frac{1}{4}$ tsp |
| | 1 egg, well beaten | |
| 200 mL | milk | $\frac{3}{4}$ cup |
| 50 mL | melted fat | $\frac{1}{4}$ cup |
| 500 mL | blueberries | 2 cups |
| 125 ml | sugar (second amount) | $\frac{1}{2}$ cup |
| 50 mL | butter | $\frac{1}{4}$ cup |
| 75 mL | flour (second amount) | $\frac{1}{3}$ cup |
| 2 mL | cinnamon | $\frac{1}{2}$ tsp |

Sift together sugar (first amount), flour (first amount), baking powder, and salt. In another bowl blend egg, milk, and melted fat. Make a well in flour mixture and add the liquid ingredients all at once. Stir quickly until flour mixture is just moistened. Pour into a medium-sized shallow glass baking dish and cover with the blueberries. For the crumb topping, cream the sugar (second amount) and butter together. Add flour (second amount) and cinnamon and mix until crumbly. Spread over the blueberries and bake at 180°C (350°F) for 40 to 50 minutes. (From Eleanor A. Ellis, *Northern Cookbook*.)

## Blueberry Crisp

| | | | |
|---|---|---|---|
| 75 mL | sugar | 1/3 | cup |
| 30 mL | cornstarch | 2 | tbsp |
| 1 mL | salt | 1/4 | tsp |
| 1 mL | cinnamon | 1/4 | tsp |
| 1 mL | nutmeg | 1/4 | tsp |
| 15 mL | lemon juice | 1 | tbsp |
| 1 L | blueberries, canned *or* frozen, sweetened to taste | 4 | cups |
| 250 mL | blueberry juice, drained from fruit | 1 | cup |
| 75 mL | butter *or* margarine | 1/3 | cup |
| 30 mL | flour | 2 | tbsp |
| 250 mL | brown sugar, firmly packed | 1 | cup |
| 750 mL | corn flakes | 3 | cups |

Combine sugar, cornstarch, salt, and spices in a medium-sized saucepan. Add lemon juice and blueberry juice and stir until smooth. Cook over low heat, stirring constantly until thickened and clear. Stir in blueberries. Pour into buttered medium-sized baking dish. To make topping, melt butter *or* margarine in heavy saucepan. Combine brown sugar and flour and add to butter. Cook, stirring constantly, over low heat for 3 minutes. Add corn flakes, mixing quickly until coated with syrup. Sprinkle evenly over blueberry mixture. Bake at 200°C (400°F) about 30 minutes, or until topping is crisp and golden brown. Serve warm or cold with cream if desired. (From Eleanor A. Ellis, *Northern Cookbook*.)

## Blueberry Slump

| | | | |
|---|---|---|---|
| 1 L | blueberries | 4 | cups |
| 125 mL | water | $^1/_2$ | cup |
| 375 mL | sugar (first amount) | $1^1/_2$ | cups |
| 5 mL | grated nutmeg | 1 | tsp |
| 250 mL | flour | 1 | cup |
| 5 mL | baking powder | 1 | tsp |
| 15 ml | sugar (second amount) | 1 | tbsp |
| 1 mL | salt | $^1/_4$ | tsp |
| | 1 egg, beaten | | |
| 45 mL | milk | 3 | tbsp |
| 30 mL | melted fat | 2 | tbsp |

Put blueberries, water, sugar (first amount), and nutmeg in a heavy saucepan and bring slowly to the boil. For the batter, sift together flour, baking powder, sugar (second amount), and salt. In a separate bowl combine the egg, milk, and melted fat, and then add to flour mixture, stirring until dry ingredients are just moistened. Drop batter mixture by spoonfuls on boiling berries. Cover and cook 10 minutes. Serve hot with cream. (From Eleanor A. Ellis, *Northern Cookbook*.)

## More for Your Interest

Vacciniums have been a favourite of native peoples all over North America. They were preserved for winter by drying, either individually like raisins, or mashed and formed into cakes. Often they were mixed with animal meat or tallow to make a sort of pemmican. The bog bilberry, common in the Arctic, was not highly esteemed by the Inuit, according to A. E. Porsild in "Edible Plants of the Arctic". They believed it caused tooth decay.

Blueberry leaves and dried fruits can be used to make a very pleasant tea. In Siberia, bog bilberries are fermented and distilled to make a strong alcoholic beverage. The French are said to use the berries to colour wine.

Wild blueberries are eaten by many types of birds and animals. Black bears are especially fond of them and will often break down entire bushes in their eagerness to get at the fruit. Care should be taken by people harvesting the berries in bear country that they do not inadvertently surprise a bear. It is best to go berry-picking with a friend and talk loudly or sing to warn animals of your presence.

# Oaks

(Beech Family)

## How to Recognize

In Canada the oaks are represented by ten native species, nine of which are full-sized trees, and one, the dwarf chinquapin oak (*Q. prinoides* Willd.), a shrub. All Canadian oaks are deciduous. The leaves are simple, alternate, and variously lobed or toothed. The male and female flowers are separate, but appear on the same tree, usually after the leaves unfold in spring. The male, pollen-bearing flowers are arranged in long, hanging clusters. The female flowers are borne singly or in small, tight groups. Oak fruits, known as acorns, are thin-shelled, oval or globe-shaped nuts, each embedded in a scaly cup.

The acorns of all oaks are edible when properly prepared, but some are much more palatable than others. The oaks of Canada separate naturally into three groups: the white oaks, the chestnut oaks, and the red oaks. The easiest way to tell the difference between these groups is by the leaves: white oak leaves have rounded lobes and teeth, without bristles at their tips; chestnut oak leaves are regularly toothed, rather than deeply lobed, and the teeth lack bristly tips; red oak leaves have sharply-pointed lobes and teeth with bristly tips. These differences are important because the acorns of the first two groups, which ripen in one season, are usually sweet-tasting, whereas those of the red oak group, which take two years to mature, are bitter and unpalatable raw, and must be leached to remove some of the strong-tasting tannic acid before they can be eaten (see Warning below).

Species of the white oak group in Canada include white oak (*Q. alba* L.), garry oak (*Q. garryana* Dougl.), and bur oak (*Q. macrocarpa* Michx.). There are four species in the chestnut oak group, including chinquapin oak (*Q. muehlenbergii* Engelm.) and chestnut oak (*Q. prinus* L.). Species of the red oak group include red oak (*Q. rubra* L.), black oak (*Q. velutina* Lam.), and pin oak (*Q. palustris* Muenchh.).

*Quercus* species
(Fagaceae)

## Where to Find

The only native oak to be found in western Canada is garry oak, which is almost entirely restricted to the southern part of Vancouver Island and to the Gulf Islands of British Columbia. The range of bur oak extends from New Brunswick to southern Manitoba and southeastern Saskatchewan. All the other oaks are restricted to various parts of southeastern Canada, mostly the Great Lakes–St. Lawrence region of southern Ontario. For further information on the characteristics and distributions of individual oak species the reader is referred to *Native Trees of Canada* by R. C. Hosie.

## How to Use

Within their range, oaks are common and usually produce a plentiful crop of acorns. Thus, although acorns are not a favourite nut with most people, they have the advantage of being available in quantity and are easy to gather from the ground in the fall. They are rich in starch and oil, and can be thoroughly enjoyable when properly treated. As already mentioned, some acorns, particularly those of the red oak group, but also of some in the other groups, such as garry oak, contain a high concentration of tannic acid, which renders them bitter and unpalatable when eaten fresh.

*Quercus garryana*

The tannin can be removed by leaching, but some of the sugars will be lost in the process. By tasting a raw acorn you can easily determine whether leaching is necessary or not. Do not eat bitter-tasting acorns in quantity without leaching (see Warning below). If the acorns are bitter, crack them open and thoroughly dry the kernels. Cover with a generous quantity of water, bring to a boil, and keep boiling for 2 hours. Pour off the now darkened water, cover the acorns with cold water, and allow them to soak for 3 to 4 days, changing the water occasionally. The acorns may now be re-dried and used whole or ground into meal. One way to remove the bitter element from the acorns without also dissolving the sugars is to add water and unflavoured gelatin to unleached ground acorns. According to Fernald and Kinsey, in *Edible Wild Plants of Eastern North America*, the gelatin removes the tannin without affecting the sugar content.

Acorns that taste sweet when fresh can be used immediately or ground to a meal, spread out on a baking sheet and dried in a slow oven, and stored for later use. Roasted and salted acorns are quite satisfactory as table nuts, tasting something like a cross between sunflower seeds and popcorn. After grinding the kernels in a blender, grinder, or with a mortar and pestle, you can cook the ground meal with water to make a breakfast mush, or substitute it for conventional flours in baking cakes, breads, and muffins. Acorn flour is also a good thickener for soups and sauces. In baking, the flour can be mixed half and half with regular wheat flour, and is especially good as a substitute or blend with cornmeal. Ground roasted acorns have been a popular coffee substitute, especially in eastern Europe. Adam Szczawinski has used this beverage frequently.

## Warning

Never eat bitter-tasting acorns in any quantity without first leaching until the bitterness has disappeared. Many cases of livestock poisoning and loss have been recorded in both North America and Europe as a result of consumption of raw acorns, especially those of the red oak group. Oak shoots and foliage are extremely poisonous, also due to tannin content. Tannins are widely found in plants, but are especially high in oaks. The medicinal use of tannins in the treatment of diarrhoea, hemorrhoids, and in the treatment of burns has been discontinued because the quantities absorbed were sufficient to cause damage to the liver. High intakes of tannin have also been implicated in some forms of cancer. For this reason one must be very careful about the identification of the oak species and the leaching procedures.

## Suggested Recipes

### Fried Acorns

| | | |
|---|---|---|
| 15 mL | cooking oil | 1 tbsp |
| 250 mL | sweet acorns | 1 cup |
| | salt to taste | |

Heat oil in heavy skillet until it just starts to smoke. Meanwhile, halve acorns, wash, and pat dry. Add to oil and sauté, stirring constantly for about 5 minutes, or until well browned on all sides. Remove, drain on paper towel, and sprinkle on salt to taste. Cool and serve as table nuts.

### Acorn–Corn Bread

| | | |
|---|---|---|
| 125 mL | acorn meal | $\frac{1}{2}$ cup |
| 125 mL | cornmeal | $\frac{1}{2}$ cup |
| 30 mL | flour | 2 tbsp |
| 10 mL | baking powder | 2 tsp |
| 2 mL | salt | $\frac{1}{2}$ tsp |
| | 1 egg | |
| 15 mL | maple syrup | 1 tbsp |
| 15 mL | cooking oil | 1 tbsp |
| 125 mL | milk | $\frac{1}{2}$ cup |

Mix together dry ingredients. In a separate bowl, beat egg, then add maple syrup, oil, and milk. Combine wet and dry ingredients with a few rapid strokes. Pour batter into a small, square, greased pan. Bake at 220 °C (425 °F) for about 20 minutes, or until bread is firm to touch. Serve hot with butter.

## More for Your Interest

Many North American Indian groups from eastern Canada to California used acorns as a staple food. They developed a number of leaching procedures to remove the bitter tannins, including mixing the kernels with wood ash and water and boiling for many hours, or burying the acorns in moist earth for several months, or powdering the kernels in a closely woven basket and setting them in a stream to be leached by the running water.

Acorns were also eaten in the past in parts of Europe, as reported by herbalist John Gerard in *The Herbal or General History of Plants* (p. 1345): "The Acorne [of the great Skarlet Oke] is esteemed of, eaten and brought into the market to be sold, in the city of Salamanca in Spaine, and in many other places of that countrey . . . . At this day in Spain the Acorne is served for a second course."

Despite the many recorded instances of acorn poisoning of livestock (see Warning above), acorns were and still are widely used, especially in Europe, to feed hogs. Gerard remarks (p. 1341): that "swine are fatted herewith [with 'oke' acorns], and by feeding hereon have their flesh hard and sound."

The oaks are in the same family as beech (*Fagus grandifolia* Ehrh.), restricted in Canada to the hardwood region from Cape Breton Island to the north shore of Georgian Bay. Beechnuts are edible and sweet in flavour, and are often used to make a good coffee substitute (see our publication, *Wild Coffee and Tea Substitutes of Canada*).

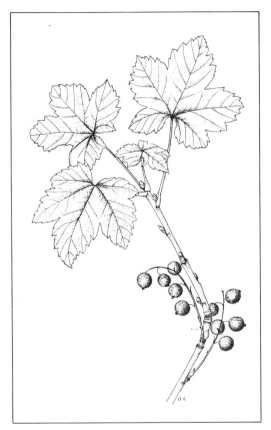

## How to Recognize

There are over ten species of wild currants across Canada, with fruits ranging in colour from red to bluish to black. Some resemble the garden varieties of currants, which were derived mainly from European species. Our native northern red currant (*R. triste* Pall.), for example, is very similar to the European red currant (*R. sativum* Syme), which is commonly cultivated in North America. Our northern black currant (*R. hudsonianum* Richards.) and American black currant (*R. americanum* Mill.) are closely related to the black garden currant (*R. nigrum* L.), also a European species. Other wild currants are quite distinctive and have no cultivated counterparts. Some, such as the red-flowering currant (*R. sanguineum* Pursh) and the golden currant (*R. aureum* Pursh), have strikingly beautiful flowers, and are grown in gardens for their ornamental value rather than for their fruit.

# Wild Currants

(Gooseberry Family)

*Ribes hudsonianum*

**Ribes species**
(Grossulariaceae)

Currants are distinguished from their close relatives the gooseberries, also in the genus *Ribes*, by the lack of spines or prickles on their stems. One notable exception is the bristly black currant [*R. lacustre* (Pers.) Poir.], which is sometimes called swamp gooseberry but is in fact one of the currants. Its stems are often profusely covered with thin, sharp spines that, with skin contact, can cause allergic reactions in some people. This species does conform to the currants in having quite small berries in large clusters, whereas most of the gooseberries have larger berries borne singly or in small clusters.

The currants are, in general, low, slender shrubs, although some can attain a height of several metres in a good location. The leaves range in size from very small, as in the squaw, or desert, currant (*R. cereum* Dougl.), to as much as 25 cm across in the western stink currant (*R. bracteosum* Dougl. ex Hook.). In general they are maple-leaf-like, with 3 to 7 pointed or rounded palmate lobes. The leaves, twigs, and

berries of many species, especially the dark-fruited ones, give off a strong, rank, skunk-like odour when touched or bruised. The flowers, usually borne in dense, drooping clusters, vary from greenish brown and inconspicuous to bright red or deep golden and very showy. The berries of some species are sweet and juicy, of others dry and insipid. None of the currants are harmful or poisonous, but some are more highly recommended than others. We think that the wild northern red currant and the various black-fruited species are superior to those with bluish fruits, although some of these are also very tasty.

**Where to Find**

Wild currants grow throughout Canada in habitats ranging from semi-arid to damp and thickly wooded. Most prefer damp, shady streambanks and moist woodlands. Several species, including the northern red currant and the northern, American, and bristly black currants, are widespread. Others, such as the western stink currant, red-flowering currant, squaw currant, and golden currant, are quite regional in their distribution. The first three are confined, in Canada, to parts of British Columbia and the golden currant to the southern prairies.

## How to Use

Currants are usually too sour or strong-tasting to be eaten raw, but some people find them refreshing when eaten fully ripe and in limited quantities. They are very rich in vitamin C and have long been used in Europe as a tonic, especially for colds and vitamin deficiencies. They make excellent jams, jellies, syrups, wines, and preserves of all kinds.

To obtain the best results in making currant preserves try to collect fruit that is firm, dry, and not over-ripe. Cook the fruit slowly, with just enough water to prevent burning, or, even better, substitute fruit juice for water. Sugar or other sweetening should be added when the fruit is cooked and tender. Currants can be frozen easily, either whole or as a purée, and we have found that red currants are delicious mixed with plums, pears, and pineapple chunks as a fancy fruit salad. This mixture can be frozen, with the addition of a little citric acid, and used whenever required. Wild currants can also be dried and used in any baking recipe calling for dried currants or raisins.

## Suggested Recipes

### Red Currant and Raspberry Jam

| | | |
|---|---|---|
| 750 mL | crushed wild red currants | 3 cups |
| 175 mL | water | $3/4$ cup |
| 1 L | raspberries (wild *or* garden) | 4 cups |
| 1.75 L | sugar | 7 cups |
| 125 mL | liquid pectin | $1/2$ cup |

Cook the red currants and water together for 10 minutes. Strain through a sieve or jelly bag. Return juice to saucepan and add raspberries and sugar. Add liquid pectin. Boil hard for 1 minute and remove from the heat. Pour into hot, sterilized jars and seal with melted paraffin. Yields about 8 medium-sized glasses. (From Eleanor A. Ellis, *Northern Cookbook.*)

## Currant – Orange Salad

| | | | |
|---|---|---|---|
| 500 mL | fresh wild currants | 2 | cups |
| | 2 large oranges, peeled and sliced | | |
| 125 mL | sugar | 1/2 | cup |
| 250 mL | boiling water | 1 | cup |
| 90 g | raspberry gelatin (2 small packages) | 6 | oz |
| 500 mL | cold water | 2 | cups |

Grind together currants and oranges. Stir in sugar, mix well, and let stand in refrigerator for a few hours. Dissolve gelatin in boiling water in a bowl, and add cold water. Stir in the orange–currant mixture, pour into mold, and let set in refrigerator. Serve with ham or pork. Makes up to 10 servings.

## Black Currant Wine (without yeast)

| | | | |
|---|---|---|---|
| 750 mL | black currants | 3 | cups |
| 500 mL | water | 2 | cups |
| 750 mL | sugar | 3 | cups |
| | a few raisins | | |

Place currants in a large jar or crock, add water, cover, and stir daily for 10 days. Then mash currants and leave for 1 week, stirring the pulp daily. After 1 week, strain through a jelly bag or several layers of cheesecloth (do not squeeze), and add sugar. Leave to ferment for 3 weeks in a warm, dark place. Strain off the liquid and bottle, adding a couple of raisins to each bottle. Cork lightly until all fermentation ceases, then tighten. Leave for at least 6 months. This is an old, Scottish recipe that produces a surprisingly good wine. For a dryer wine, try reducing the proportion of sugar.

## Black Currant Concentrate

Wash any quantity of black currants and place in a saucepan over medium heat with just enough water to prevent burning. Stir gently until soft and pulpy, pressing with a wooden spoon. Rub the pulp through a sieve and return to saucepan. Sweeten to taste with sugar or honey, and pour at once into hot, sterilized jars. Seal and store in refrigerator until needed. This concentrate makes delicious drinks when diluted with cold or hot water, ginger ale, or other soft drinks. It is very rich in vitamin C and can be taken as a tonic throughout the winter months.

## More for Your Interest

The genus name *Ribes* is said to have originated from the old Danish colloquial name *ribs* for red currant. The term *currant* was originally applied to a type of small black grape imported from Greece, whose raisins were used in baking. They are now raised mainly in California under the name of black corinth grape or zante currant.

True currants, especially black currants, are widely cultivated and used in many commercial preparations, such as syrups, jellies, cough preparations, and throat lozenges. In North America, however, the planting of currants has been discouraged or even prohibited in some regions, as they and the gooseberries are alternate hosts of the white pine blister rust disease that has afflicted both eastern and western pine species. Fortunately, recent varieties of pine developed by the Canadian Government are immune to this disease, and now the cultivation of currants is steadily increasing in Canada.

The currants were a common food of the Indian peoples throughout Canada, often eaten with oil, fish eggs, meat, or other fruits. The Kwakiutl Indians of British Columbia, for example, boiled stink currants with salal berries in tall cedarwood boxes and dried them for four or five days. In winter the dried berry cakes were soaked in water overnight, then eaten mixed with eulachon grease (rendered from a small, oily fish of that name) at feasts. The guests had to eat all the currants in their dishes or they would have bad luck. The Thompson Indians of British Columbia believed that black currant (*R. hudsonianum*) growing around the edge of a lake was a sure indication that the lake contained fish.

# Wild Gooseberries

(Gooseberry Family)

## How to Recognize

Gooseberries are usually placed by botanists in the same genus as the currants (*Ribes*), although some prefer to designate them as a separate genus, *Grossularia*. Their sharp spines or thorns on the twigs easily distinguish them from the currants, which are unarmed. Also, currants are usually borne in large clusters, whereas gooseberries are borne singly or in small clusters.

Wild gooseberries closely resemble the cultivated forms found in gardens. All the gooseberry species found in Canada are deciduous shrubs about 1 to 2 m tall, with palmately lobed (maple-leaf-like) leaves 2 to 5 cm wide, and with spiny branches, often with 2 types of prickles—small, slender ones spaced along the twigs, and stouter, longer thorns at the leaf nodes. The flowers are generally small, usually with reddish, reflexed sepals and protruding stamens. The many-seeded berries are variable in size and are either smooth, hairy, or bristly. Some are reddish, some purplish, and some black, but all are juicy, with translucent skins, and most are quite tasty when ripe. The berries can be recognized by the brownish "wick" at the end of each, actually the remains of the flower calyx.

One of the best-flavoured gooseberries is the western species *R. divaricatum* Dougl. It is sometimes called "Worcesterberry" because it seems to have been first sold as a commercial fruit by a nurseryman in Worcester, who thought it was a black currant–gooseberry hybrid. The purplish-black berries are numerous and the bushes are said to be immune to mildew. According to *Sturtevant's Edible Plants of the World*, nineteenth-century botanist John Lindley stated that, of all the species he observed during his travels in America, this species had the largest and finest flavoured fruits.

Other gooseberry species common in parts of Canada include hairy gooseberry (*R. hirtellum* Michx.), smooth gooseberry (*R. oxyacanthoides* L.), bristly gooseberry (*R. setosum* Lindl.), black gooseberry (*R. irriguum* Dougl.), and the bristly-fruited dog bramble, or dogberry (*R. cynosbati* L.). One species, *R. lobbii* Gray, restricted in Canada to southwestern British Columbia, is exceptional because its flowers are deep crimson, large, and very showy. It is commonly known as sticky, or gummy, gooseberry because its fruits are covered with sticky hairs. When ripe they are sweet and very large, almost the size of cultivated gooseberries. The fruits of all the above-mentioned species are wine-red or various shades of purple when ripe.

*Ribes* species
(Grossulariaceae)

*Ribes lobbii*
*Ribes divericatum*

*Ribes hirtellum*

## Where to Find

Gooseberries are found almost everywhere in Canada except in the Far North, in areas ranging from moist woods to dry, rocky uplands. *R. divaricatum* is restricted to western British Columbia, *R. hirtellum* occurs from Manitoba to Newfoundland, *R. oxyacanthoides* from British Columbia to northern Ontario, *R. setosum* across the southern prairies, *R. irriguum* in central British Columbia, and *R. cynosbati* in southern Quebec and Ontario.

## How to Use

The fruits of all our wild gooseberry species are edible in both raw and cooked form. Although some species have fruit with sticky, bristly skin or a rather strong odour, none are poisonous and all of them can be used in jellies, jams, and other preserves. *R. divaricatum* and *R. cynosbati* fruits are especially good when ripe. They are delicious in tarts and pies and can be used in the same way you would the cultivated gooseberries. They should be picked when fully ripe, however, as they are apt to be quite tart. Any of the wild gooseberries can be used in the following recipes.

## Suggested Recipes

### Gooseberry Jam

| | | |
|---|---|---|
| 2 kg | gooseberries | 4 lb |
| 250 mL | rhubarb juice (from stewed rhubarb) | 1 cup |
| | 1 bunch of elder flowers (optional) | |
| 2 kg | sugar | 4 lb |

Clean and wash berries. Put into large saucepan with rhubarb juice and elder flowers (for flavouring). Cook until berries are soft (10 to 15 minutes). Add sugar, stir until boiling, and boil for an additional 10 to 15 minutes, or until jam reaches jelling stage. Remove elder flowers and pour jam into hot, sterilized glasses, seal with melted paraffin, and store in a cool place. Makes about 10 medium-sized jelly glasses.

### Gooseberry Tarts

Line tart shells with plain pastry, prick with a fork, and bake in a 200°C (400°F) oven for about 10 minutes, or until lightly browned. Fill with gooseberry jam (see previous recipe), top with meringue, and brown in 180°C (350°F) oven.

## Gooseberry–Milk Jelly Dessert

| | | |
|---|---|---|
| 500 mL | milk | 2 cups |
| 50 mL | sugar | ¼ cup |
| 30 g | powdered gelatin | 1 oz |
| 500 mL | gooseberries, stewed and sieved | 2 cups |
| | a few drops of lemon flavouring | |

Bring milk to boiling point and pour it over gelatin and sugar in a bowl, stirring until thoroughly dissolved. Stir from time to time until mixture is cold and beginning to thicken, then fold in stewed gooseberries and flavouring. Pour into small, individual moulds and leave to set. Serve cold as a dessert. Serves 4–6.

## Gooseberry Preserves

| | | |
|---|---|---|
| 1 L | gooseberries | 4 cups |
| 750 mL | sugar | 3 cups |
| 125 mL | pineapple juice | ½ cup |

Wash and clean gooseberries and place in large saucepan. Add pineapple juice and bring to the boil, stirring well. When boiling add sugar. Boil until thick and clear (15 to 20 minutes), stirring constantly. Remove from heat, cool, and pour into hot, sterilized jars. This is a good, tart preserve, to serve with meats and cottage cheese. Yields about 5 medium-sized glasses.

## More for Your Interest

Gooseberries are widely cultivated in England and are quite popular in parts of North America as well. Hundreds of varieties have been developed from both New World and Old World species.

Wild gooseberries were commonly eaten by native peoples in Canada, fresh or occasionally dried. Sometimes they were mixed with other fruits or with various types of edible roots for drying. The Bella Coola Indians of coastal British Columbia used to enjoy a sauce made from the green, unripe berries of *R. divaricatum* mixed with a few leaves from the same bush.

Gooseberry thorns were often used by Indians as probes for boils, for removing splinters, and for tattooing. Some people made fish-hooks from them. The Straits Salish Indians of Vancouver Island used gooseberry roots to bind reef-nets, used in salmon fishing.

**113**

# Hickories

(Walnut Family)

## How to Recognize

There are six species of hickory to be found in Canada, all very much restricted in distribution. They are medium-sized, deciduous trees with deep root systems and straight trunks with open, spreading crowns. The bark is smooth and grey when young, becoming scaly or rough with age. The leaves, like those of the related walnuts, are pinnately compound, with 5 to 11 oval-shaped, finely toothed, pointed leaflets. Male and female flowers are borne in separate clusters on the same tree. The female flowers ripen into edible nuts, each enclosed in a semi-woody husk that splits open partially or wholly at maturity. The shell of the nut, unlike that of the walnut, is relatively smooth and not deeply grooved.

The most important species for edible nuts is the shagbark hickory [*C. ovata* (Mill.) K. Koch]. As the name implies, the bark of this species is shaggy, peeling off in long, thin strips that give the trunk an untidy appearance. There are usually 5 (sometimes 7) leaflets per leaf, and the husk of the nut is thick and woody, splitting to the base at maturity. The shell is quite thick and hard. The other edible-fruited species of hickory occurring in Canada are the big shellbark hickory [*C. laciniosa* (Michx. f.) Loud.], also with shaggy bark and thick husks and shells but with usually 7 leaflets per leaf; and the mockernut hickory (*C. tomentosa* Nutt.), with thin, furrowed bark and thinner husks than the two previous species. Less important are the pignut hickory [*C. glabra* (Mill.) Sweet], the red hickory [*C. ovalis* (Wang.) Sarg.], often considered a form of the pignut hickory, and the bitternut hickory [*C. cordiformis* (Wang.) K. Koch]. All of these species are described in detail by R. C. Hosie in *Native Trees of Canada*. Some are quite variable and tend to hybridize, making identification extremely difficult in some cases.

*Carya* species
(Juglandaceae)

## Where to Find

The shagbark hickory is found in Canada in rich, moist soils in the southernmost parts of Ontario, especially north from Lake Erie, and northeastwards into Quebec along the St. Lawrence Valley. The distribution of the bitternut hickory is similar. The other species are more restricted, occurring only in some localities within the deciduous forest region around Lake Erie, usually with other hardwoods. All of these species are more widespread in the United States. The shagbark hickory, for example, extends throughout the eastern United States from Texas and Nebraska to the Atlantic Ocean.

## How to Use

The hickories are among the most important of our native nuts. The well-known pecan nut of the eastern United States is actually a type of hickory [*C. illinoensis* (Wang.) K. Koch], but, unfortunately, its closest relative in Canada, the bitternut hickory, is not at all comparable, being too bitter to eat. The other species, however, particularly the

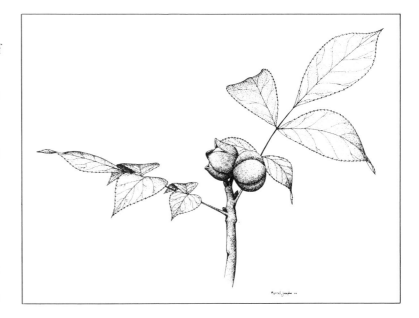

shagbark, big shellbark, and mockernut hickories, are edible and lend themselves to a wide variety of culinary applications. You can simply crack them with a hammer or stone and eat them raw, or use them in making all types of desserts and confections. Try them in any recipe calling for walnuts or pecans; they give a different and, in the opinion of many, superior taste.

Gather hickories after they have fallen to the ground in the fall, or even the following spring. They are also relished by squirrels, birds, and domesticated hogs, however, and sometimes it is hard to beat the competition. The nuts are a good source of protein, carbohydrate, fat, and iron, and are an excellent source of phosphorus.

## Suggested Recipes

### Hickory–Cranberry Mince Pie

|         |                               |         |
|---------|-------------------------------|---------|
|         | pastry for 2-crust *or* lattice-top pie |         |
| 175 mL  | hickory nuts                  | 3/4 cup |
| 125 mL  | raisins                       | 1/2 cup |
| 250 mL  | raw cranberries               | 1 cup   |
|         | 2 medium apples, cored and sliced |     |
| 15 mL   | flour                         | 1 tbsp  |
| 5 mL    | cinnamon                      | 1 tsp   |
| 2 mL    | nutmeg                        | 1/2 tsp |
| 250 mL  | sugar                         | 1 cup   |
| 50 mL   | boiling water                 | 1/4 cup |
| 15 mL   | butter                        | 1 tbsp  |
|         | whipped cream *or* ice cream (optional) |   |

Line pie plate with bottom pastry crust. Chop *separately* in a blender the nuts, raisins, cranberries, and apples. Mix in a bowl with the remaining dry ingredients and add the boiling water last. Place in pie shell, dot with butter, and cover with top crust *or* lattice. Bake at 220°C (425°F) for 15 minutes, then lower temperature to 190°C (375°F) and bake for another 25 to 30 minutes, or until crust is well browned. Serve with whipped cream *or* ice cream if desired.

## Hickory-Nut Torte

|         |                                          |          |
| ------- | ---------------------------------------- | -------- |
|         | 7 medium (or 8 small) egg yolks          |          |
| 250 mL  | sugar                                    | 1 cup    |
|         | 7 medium (or 8 small) egg whites         |          |
| 750 mL  | hickory nuts, well ground                | 3 cups   |
| 10 mL   | brandy (first amount)                    | 2 tsp    |
| 50 mL   | softened butter                          | 1/4 cup  |
| 250 mL  | icing sugar                              | 1 cup    |
| 50 mL   | strong, hot coffee                       | 1/4 cup  |
| 15 mL   | brandy (second amount)                   | 1 tbsp   |
| 2 mL    | vanilla flavouring (optional)            | 1/2 tsp  |

Cream egg yolks and sugar together. In a separate bowl, beat egg whites until stiff and fold into first mixture. Mix in nuts and brandy (first amount) and pour the mixture into an angel-cake tin. Bake in slow oven at 140°C (275°F) for about 1 hour, or until torte springs back when touched. Cool on a wire rack. Meanwhile, make mocha icing: cream together butter and icing sugar and gradually blend in coffee, brandy (second amount), and flavouring if desired. Mix thoroughly until thick enough to spread, and cover torte when cool. Serves 6–8.

## Hickory Cookies

|         |                      |          |
| ------- | -------------------- | -------- |
| 125 mL  | softened butter      | 1/2 cup  |
| 250 mL  | brown sugar          | 1 cup    |
| 175 mL  | hickory nut pieces   | 3/4 cup  |
|         | 1 egg, beaten        |          |
| 250 mL  | flour                | 1 cup    |
| 2 mL    | vanilla              | 1/2 tsp  |
|         | a few hickory halves |          |

Cream butter and sugar. Add nuts, egg, flour, and vanilla, and mix well. Drop in small spoonfuls on a greased cookie sheet (space 3 cm apart) and place half a hickory nut on top of each cookie. Bake 10 to 12 minutes at 180°C (350°F). Yields about 2 dozen small cookies.

## Hickory Christmas Balls

| | | | |
|---|---|---|---|
| 500 mL | hickory nuts | 2 | cups |
| 500 mL | dates | 2 | cups |
| 50 mL | honey | ¹/₄ | cup |
| 30 mL | rum | 2 | tbsp |

Put nuts and dates separately through food chopper or blender. Mix together chopped dates and half of the nuts. Add the honey and rum to this mixture. Stir well and form into small balls. Roll balls in remainder of chopped nuts. Chill in fridge and serve when cold. These are easy to make and always welcome in the Christmas season. Makes 2 to 3 dozen.

### More for Your Interest

According to Euell Gibbons, in *Stalking the Wild Asparagus*, the name *hickory* is of Indian origin, being derived from the term *powchicora*, applied in at least one native language to a kind of nut-milk made by pounding the unshelled nuts and boiling them in water. This milky, oily liquid was used as a food and condiment by many Indian groups. The early European settlers also learned to make use of hickory oil and "milk". The sweet sap of the hickory trees was tapped in the spring in the same manner as that of the sugar maple.

Hickory wood is extremely heavy, hard, and strong, though not very durable. It is used for making tool handles, as well as wheel spokes, bows, skis, and other items requiring great strength. The early immigrants to the West frequently carried hickory wood on their "prairie schooners" to mend wagon wheels, ox yokes, and other wooden parts for their wagons.

The wood and bark of hickory, especially of the bitternut hickory, are famous for their use in smoking meats. Hickory-smoked bacon and ham are sold throughout this continent, and hickory-barbecued steak is one of the best-flavoured to be had anywhere. Unfortunately, some hickories have been so exploited for their valuable wood that they are in danger of becoming extinct in some areas.

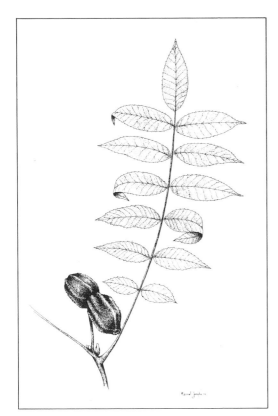

*Juglans cinerea*

## Other Names
Black walnut is also known as American, or American black, walnut; butternut is sometimes called white walnut, or oilnut.

## How to Recognize
There are about fifteen species of walnut distributed throughout the world. Of these only two, the black walnut and butternut, are native to Canada, although the well-known English walnut (*J. regia* L.), source of the commercial walnut, is widely planted as an ornamental in Canada.

Black walnut and butternut are, like the English walnut, small to medium-sized trees (black walnut is somewhat larger than butternut), with coarse, spreading branches and rough, furrowed bark. The leaves are large, alternate, and pinnately compound, with many pointed, elliptical leaflets (15 to 23 in black walnut, 11 to 17 in butternut). The terminal leaflet of black walnut is much smaller than the laterial leaflets, or often lacking. The terminal leaflet of butternut is relatively large and conspicuous. The flowers are unisexual, the male and female being separate but both on the same tree. The male flowers are borne in drooping catkins and the female in short, few-flowered clusters. The female flowers ripen into large, edible, hard-shelled nuts surrounded by a thick rind, or hull. Black

# Black Walnut and Butternut
(Walnut Family)

walnuts are spherical with hairy but not sticky hulls, and shells deeply grooved but not jagged. Butternuts are about twice as long as broad, with densely hairy, sticky hulls and grooved, jagged shells.

### Where to Find
Black walnut occurs naturally only in the warmest, most southern part of Ontario, in deep, rich soil with other deciduous hardwoods. It is often planted as an ornamental, however, and can be found under cultivation considerably north of its natural range. Butternut is found in a variety of sites from southeastern Ontario to southern Quebec along the St. Lawrence Valley and in the St. John's River Valley of New Brunswick.

### How to Use
Black walnuts and butternuts are rich and flavourful. Although black walnuts are sometimes difficult to extract from the shells, they are well worth the effort. They can be used in the same way as English walnuts, and many people, including us, prefer them to English walnuts for eating. They are used commercially in ice cream, and there is nothing better than a dish of homemade ice cream flavoured with maple syrup and bits of black walnut. Try black walnuts instead of pecans as a topping for butter tarts or apple puddings.

Butternuts have hard shells, but the kernels, sweet but very oily, come out easily. In the past they were gathered for home use and were often sold in the markets. You can extract the oil by mincing or pressing the kernels and simmering them in a little water; the oil will float to the top and can be skimmed off and used on salads. The soft, green nuts can be pickled with vinegar, sugar, and spices in the same manner as English walnuts. It is said that if the nuts are soft enough for a knitting needle to pass through the husk and nut without hindrance they are at the right stage for pickling. Before being pickled, the nuts should be scalded and the thick fuzz rubbed off the outside of the hulls. Pickled butternuts make an excellent relish with meats in winter.

Walnuts and butternuts should be stored in a cool, dry place, as they tend to become rancid quickly at warm temperatures.

## Warning

It is advisable to use walnuts and butternuts in moderation because, according to Lewis and Elvin-Lewis in *Medical Botany*, they contain a cell-modifying chemical (mitogen), which can have harmful effects on the blood. The leaves, bark, and nuts, including the hull, of black walnut have been widely used in Europe for the treatment of fungal skin infections. Teas made of the bark have a cathartic and laxative effect.

## Suggested Recipes

### Walnut–Honey Balls

| | | |
|---|---|---|
| 500 mL | black walnuts *or* butternuts | 2 cups |
| 500 mL | dates | 2 cups |
| 50 mL | honey | 1/4 cup |
| 15 mL | lemon juice | 1 tbsp |
| 15 mL | rum *or* brandy | 1 tbsp |

Chop dates and walnuts *or* butternuts separately in food chopper. Mix together dates and one half of the nuts. Moisten the mixture with honey, lemon juice, and rum *or* brandy. Form into small balls and roll them in the remainder of chopped nuts. Makes 2 to 3 dozen.

### Cinnamon Nuts

| | | |
|---|---|---|
| 250 mL | black walnuts *or* butternuts | 1 cup |
| 250 mL | blanched almonds | 1 cup |
| | 2 egg whites | |
| 125 mL | sugar | 1/2 cup |
| 30 mL | cinnamon | 2 tbsp |

Combine the nuts in a bowl. Add egg whites, beaten slightly, and work between the fingers until all the nuts are coated. Add sugar and cinnamon, mixed together, and stir until all the nuts are coated. Shake the nuts to remove excess sugar. Spread on an oiled baking sheet and bake at 150°C (300°F) for 10 to 15 minutes.

### Salted Nuts

| | | |
|---|---|---|
| 500 mL | black walnuts *or* butternuts | 2 cups |
| 15 mL | vegetable oil | 1 tbsp |
| 5 mL | salt | 1 tsp |

Place nuts in a bowl, add oil, and toss lightly. Spread on a baking sheet, sprinkle with salt, and bake at 150°C (300°F) for 10 minutes, stirring from time to time.

## Hickory-Smoked Nuts

| 250 mL | black walnuts *or* butternuts | 1 cup |
|---|---|---|
| 30 mL | melted butter | 2 tbsp |
| 2 mL | salt | 1/2 tsp |

Brush walnuts *or* butternuts with melted butter. Salt and place on a sheet of aluminum foil over hickory smoke for 20 minutes. Hickory nuts are our favourite nuts to serve with drinks and can be made when you are barbecuing meat or salmon. If you are in a hurry and must settle for second best, try sprinkling the buttered nuts with hickory-flavoured salt, and roasting them on a baking sheet in a 150°C (300°F) oven for 20 minutes.

## More for Your Interest

Many Indian peoples enjoyed black walnuts and butternuts in a variety of ways. The Iroquois crushed the kernels and mixed them with cornmeal, beans, and berries to make bread. They also extracted the oil to use as a seasoning in bread, pumpkin, squash, and other foods, and as a hair oil, either alone or mixed with bear grease. The crushed nuts and pulp left from oil-making were added to hominy and corn soup to make it rich.

Walnut and butternut trees can be tapped like sugar maples to yield a sweet sap that can be rendered into syrup.

The husks of black walnut can be used, without a mordant, to make a dark-brown or black dye, one of the oldest home dyes in North America. Butternut hulls make a light or dark-tan dye. During the American Civil War many of the Confederate soldiers' uniforms were dyed at home with butternut hulls, and the colour of the uniforms came to be known as "butternut".

The wood of black walnut is strong and hard, with a fine decorative grain and attractive surface when polished. Both it and the softer butternut wood are used extensively for interior work, furniture, and boat-building.

## How to Recognize

A short-trunked deciduous tree with a dense, rounded crown, this mulberry seldom exceeds 10 m in height. The bark is reddish brown, separating in long, flaky plates. The yellowish-green leaf blades are relatively large, up to 12 cm long, and widest below the middle. They are simple with pointed tips and toothed margins, unlobed or variously lobed. The leaf stalks are long and slender and divide into 3 prominent veins at the base of the blade. Male and female flowers are borne in separate clusters or occasionally in mixed clusters, appearing with or before the leaves. The fruits are compact aggregates resembling blackberries, about 2.5 cm long, dark red to blackish, juicy, and flavourful.

Red mulberry is the only native mulberry in Canada, but another species, the white mulberry (*M. alba* L.), originating in eastern Asia, can also be found in some parts of the country. The leaves are more lustrous than those of red mulberry and are smooth underneath, whereas those of red mulberry are hairy. The fruit is usually whitish or reddish.

## Where to Find

Red mulberry occurs naturally only in the deciduous forest region in the southernmost part of Ontario. However, it has been planted as an ornamental and for its fruit in other parts of southern Canada, and has escaped from cultivation in parts of southern Ontario and British Columbia. It prefers moist, rich soils and usually grows mixed with other hardwoods. White mulberry is also a common escape in some areas of Ontario.

## How to Use

Mulberry fruits are juicy and sweet when ripe. They are delicious raw, in fruit beverages, or in baked desserts. Once highly valued by native peoples and white settlers, they have been unjustly neglected in recent years. They are very easy to recognize, as they are the only blackberry-like fruit growing on a tree. Often they are very abundant. Because mulberries do not ripen all at the same time the fruit of a single tree can be enjoyed over a period of several weeks. To harvest the ripe fruits, which drop off very easily, spread a sheet on the ground beneath the tree and shake the branches gently.

# Red Mulberry

(Mulberry Family)

*Morus rubra* **L.**
(Moraceae)

Some people like to combine mulberries with more acid fruits, such as gooseberries or cherries, but this is a matter of individual taste. For pies, juice, or other uses, gather the berries when fully ripe for the best flavour and bouquet. The berries can easily be frozen or dried like raisins and used in muffins and fruit cakes.

## Warning

Be careful not to eat the raw fruit until it is ripe. Unripe fruits and the milky sap in the leaves and stems of mulberries are toxic and can cause gastric upsets. The leaves and stems also may cause dermatitis if touched by susceptible individuals.

## Suggested Recipes

### Mulberry Fudge

| 150 mL | ripe mulberry juice | 2/3 cup |
|---|---|---|
| 500 mL | sugar | 2 cups |
| 30 mL | butter | 2 tbsp |

Cook about 375 mL (1½ cups) of mulberries lightly, mash and drain through a fine sieve or jelly bag to obtain the juice. Mix juice together with sugar and butter and place over low heat until sugar is dissolved. Bring to a boil on medium heat and boil *without stirring* until soft-ball stage on a candy thermometer is reached (115°C or 240°F). Remove from heat. Cool to lukewarm and beat with a wooden spoon until mixture loses its gloss. Press into a buttered pan, and cut into squares before the candy hardens. Keep in a tightly covered dish or freeze because this candy dries quickly when exposed to the air.

In place of mulberry, you can use other wild fruit juices that are low in pectin, such as saskatoon berry, blueberry, elderberry, raspberry, ripe blackberry, or may-apple. (From Constance Conrader, "Wild Harvest".)

### Mulberry Pie

| | pastry for 2-crust pie | |
|---|---|---|
| 750 mL | ripe mulberries | 3 cups |
| 250 mL | sugar | 1 cup |
| 50 mL | flour | ¼ cup |
| 2 mL | salt | ½ tsp |
| 25 mL | butter | 1½ tbsp |

Line the bottom of a medium-sized pie plate with pastry, and prick with a fork. Fill with washed, de-stemmed mulberries. Mix together sugar, flour, and salt, and sprinkle over top. (Use more or less sugar, to your own taste.) Dot with butter, cover with top crust, and seal the edges with a little water. Prick top in a pattern to allow steam to escape, and bake at 200°C (400°F) for about 40 minutes, or until crust is golden brown.

## Mulberry Jelly

| | | |
|---|---|---|
| 1 L | ripe mulberries | 1 qt |
| 1 L | slightly unripe mulberries | 1 qt |
| 500 mL | water | 2 cups |
| | sugar | |

Combine berries and water in a large saucepan and simmer over medium heat, stirring frequently, until berries are soft. Strain through a jelly bag overnight, or until dripping stops. Measure juice and return to saucepan with an equal volume of sugar. Bring to a boil, and boil, stirring constantly, for about 5 minutes, or until jelly stage is reached. Pour into hot, sterilized jelly glasses and seal with melted paraffin. Yields about 8 medium-sized jelly glasses.

## More for Your Interest

A number of economically important plants, including figs (*Fiscus*), bread-fruit (*Artocarpus*), hops (*Humulus*), and hemp (*Cannabis*), are in the mulberry family.

Mulberry wood is heavy, hard, and straight-grained. It is little used commercially but its durability makes it valuable for making posts and barrels, and for boatbuilding.

Since white mulberry foliage is a major food source of the silkworm, the white mulberry has been cultivated in Japan and China for many centuries for use in silk culture as well as for its edible fruits. It was introduced to North America more than two hundred years ago in an attempt to initiate a silk industry in the New England states, but this venture proved unprofitable. White mulberry established itself rapidly and has since spread over much of eastern North America.

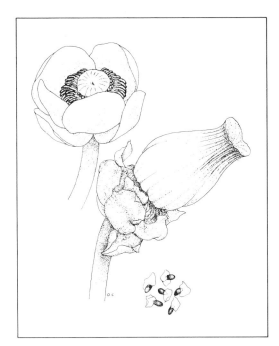

## Other Names
Yellow pond-lily is also called yellow water-lily or spatterdock; lotus-lily is also known as water chinquapin, American water-lotus, pond-nut, or yellow nelumbo.

## How to Recognize
Yellow pond-lily is easy to recognize as it usually grows along the edges of lakes and ponds, with its leaves floating on the surface, attached by long, cord-like stems to the fleshy rhizomes buried in the mud at the bottom. Occasionally the plants are seen growing in damp mud, rather than submerged in water. Here the leaf stems are shorter and the leaves more upright. Pond-lily leaf blades are leathery and heart-shaped, with the stem attached to the indented base. The leaf blades vary considerably in size, but some can reach a length of almost half a metre. The flowers are large and showy and seem almost artificial. There are 2 sets of outer floral bracts, or sepals; one is leathery and greenish, the other waxy and bright yellow (rarely reddish-tinged). The inner petals are small and inconspicuous. Inside are numerous reddish to purplish stamens surrounding a stout, flaring pistil. As the fruit matures, the petals and sepals rot away, leaving a large, fleshy, green capsule filled with numerous seeds. Eventually the pulpy flesh of the fruit disintegrates to release the edible seeds.

# Yellow Pond-Lily and Lotus-Lily
(Water-Lily Family)

***Nuphar polysepalum* Engelm. and *Nelumbo lutea* (Willd.) Pers.** (Nymphaeaceae)

Three other species of *Nuphar* occur in Canada, and all are similar in appearance, but the flowers and fruits are smaller than those of *N. polysepalum*. Their seeds can be eaten in the same way.

The lotus-lily is also aquatic, with large, round leaves, some half a metre or more across. They are usually raised above the surface of the water by the long, centrally-attached stalk. The flowers, also elevated above the water, are large, solitary, and pale yellow, with numerous sepals and petals. The seeds are nut-like and about the size of acorns or hazelnuts. They are embedded in the surface of a top-shaped receptacle about 10 cm across, which is perforated on the upper surface where the seeds are located.

## Where to Find

Species of yellow pond-lily are found in lakes, ponds, and swampy places all across Canada, but the large-fruited *N. polysepalum* occurs, in Canada, only in British Columbia and Alberta. Lotus-lily is quite rare and is restricted in Canada to ponds and lake-margins in the southern Great Lakes region of Ontario. Its range extends southwards to tropical America.

## How to Use

Pond-lily seeds, extracted from the mature pods, are nutritious and of good flavour. We recommend that you prepare them as the Indians did, by parching them in a hot frying pan until they swell and pop open slightly, somewhat like popcorn. These cracked seeds can be eaten as is, or pounded into meal and used to make bread and porridge, or for thickening soups.

Lotus-lily seeds can be easily extracted from the large, flat-topped receptacles. When the seeds are immature the outer skin is fairly soft and easy to remove. The seeds can be eaten at this stage, raw or cooked, and some people say they resemble chestnuts in flavour and texture. In fact, the alternate name, water chinquapin, comes from the similarity of the seeds to the chinquapin (or chinkapin) chestnuts of the southern United States.

When the seeds are fully ripe, the outer shell is hard and thick, and the nuts must be soaked in hot water or thoroughly parched over a fire to loosen the kernel. They can then be crushed and the bits of hard shell removed by hand-sorting or winnowing. The ripe inner kernels can be eaten without further preparation or can be boiled, baked, or roasted, and ground into flour for making bread or thickening soups.

Both lotus-lily and pond-lily seeds are easiest to gather from a boat or canoe. Sometimes it is possible to gather lotus-lily seeds in wintertime in areas where there is not much snow cover and where the lakes are frozen and easy to walk on, since the fruits are raised up from the surface.

Both types of seed, once extracted and dried, can be stored in a cool, dry place for considerable periods of time, and both are an excellent survival food for campers and hikers.

## Suggested Recipes

### Parched Pond-Lily Seeds

| 25 mL | vegetable oil | 1½ tbsp |
|---|---|---|
| 500 mL | pond-lily seeds, extracted from ripe fruit | 2 cups |
| | salt to taste | |

In a cast-iron frying pan or heavy skillet heat the oil until it starts to bubble. Add the pond-lily seeds, which have been cleaned of the fruit pulp and sun-dried. Cook slowly, shaking the pan continuously or stirring constantly to prevent seeds from burning. The seeds will start to swell and the outer skins will crack open. You can sprinkle salt over them and eat them as a snack, or grind them to a meal in a food grinder or between two flat stones. These seeds are easy to prepare over a campfire.

### Pond-Lily Seed Porridge

| 750 mL | water | 3 cups |
|---|---|---|
| 5 mL | salt | 1 tsp |
| 250 mL | pond-lily seed meal | 1 cup |

Add salt to water and bring to a rolling boil. Stir in meal (see previous recipe for preparation) and boil hard for about 15 minutes, stirring constantly, then reduce heat and cook slowly for about 1 hour, stirring occasionally to keep porridge from sticking. Serve hot at breakfast with raisins or berries, milk, and sugar or honey if desired. This makes an excellent camping breakfast. Any leftovers can be cooled, sliced up, and fried in bacon grease for the evening meal. Serves 2.

### Roasted Lotus Nuts

| 250 mL | lotus-lily seeds | 1 cup |
|---|---|---|
| 15 mL | melted butter | 1 tbsp |
| 2 mL | salt | 1/2 tsp |

Soak lotus-lily seeds in hot water and remove outer shells. Mix seeds with melted butter and salt, spread on a cookie sheet and roast in a moderate oven at about 150°C (300°F) for 1 hour, or until kernels are well browned. Serve hot or cold as a snack.

### Lotus Nuts with Rice, Chinese Style

| 125 mL | lotus-lily seeds | 1/2 cup |
|---|---|---|
| 1 L | water | 4 cups |
| 250 mL | uncooked rice | 1 cup |
| 30 mL | sugar | 2 tbsp |
| 2 mL | ginger powder | 1/2 tsp |

Soak lotus-lily seeds in hot water and remove outer shells. Wash and place in saucepan with the water. Bring to a boil, then reduce heat, and simmer gently for 1 hour. Wash rice, add it to the lotus-lily seeds, and simmer together until rice is soft. Mix in sugar and ginger and serve hot as a pudding. (This dish is best when made with special glutinous rice, available in some Chinese markets.) Serves 4.

**More for Your Interest**

Pond-lily seeds were a staple food of the Indians of California, particularly the Klamath Indians. They spent many days each summer gathering the ripe pods from the Klamath Marsh, which was once estimated to contain at least 4000 hectares (10,000 acres) of pond-lilies, or *wokas*, as they were called. The pods were harvested in large sacks and allowed to dry for a few days. The seeds were then extracted and stored for later use.

Many books on edible plants state that the thick, fleshy rhizomes of the yellow pond-lily are edible and nutritious. However, we have tried preparing them in many different ways and have always found them bitter and unpalatable. We know of others who have had a similar experience with them. The tuberous rootstocks of the lotus-lily, on the other hand, are apparently very pleasant and agreeable to eat when boiled, being described by some as only slightly inferior to sweet potatoes. In addition, the young leaf stalks and shoots of the lotus-lily are said to make an acceptable potherb.

Neither of these plants is a true lily, but both are related to the ornamental water-lilies that have been introduced throughout North America.

The lotus-lily of North America is closely related to the famous Oriental sacred lotus (*Nelumbo nucifera* Gaertn.), whose seeds and rootstocks are an important food in southeastern Asia. This species, with large pink flowers, has become locally established through cultivation in parts of North America. Both seeds and rootstocks are commonly sold in the Chinese markets of North American cities.

# Saskatoon Berries

(Rose Family)

## Other Names
Serviceberry, sarvisberry, Junebush, shad-bush, sugar pear, Indian pear, grape-pear.

## How to Recognize
There are at least fifteen different species of saskatoon berry in Canada. Of these, the most abundant is undoubtedly the highly variable western species, *Amelanchier alnifolia* Nutt. Other significant species, all eastern in distribution, include *A. canadensis* (L.) Medic., *A. arborea* (Michx.) Fern., *A. stolonifera* Wieg., *A. sanguinea* (Pursh) DC., and *A. bartramiana* (Tausch) Roemer. All have sweet, edible fruits.

The saskatoons are deciduous and range in size from low, straggly shrubs to small, bushy trees 5 m or more tall. The bark is greyish to red, and smooth. The leaves are simple, oval to elliptical, round-tipped to pointed, and usually finely to coarsely toothed around the margins. The flowers are white, 5-petalled, and showy, in tight to loose clusters, depending on the species. Blooming time ranges from March to June, often before the leaves expand. The fruits are berry-like pomes, each containing 5 to 10 whitish seeds. They resemble large blueberries, being fleshy and dark blue to blackish in colour, with sweet, juicy pulp.

## Where to Find
Saskatoons are found in every province in Canada, as well as in the Yukon and Northwest Territories. They grow in thickets, open woods, bogs, gulleys, and on stream banks and dry slopes, from sea level to subalpine elevations. *A. alnifolia* ranges from western Ontario to British Columbia and the Yukon; the other species occur in various parts of eastern Canada, from Newfoundland to Ontario.

## How to Use
The fruits of all species of *Amelanchier* are edible, and most are juicy and pleasant. However, the taste, seediness, and juiciness of the fruits vary with the different species, variety, habitat, and local weather conditions. If you have tried saskatoons in one area and found them unsatisfactory, do not give up; the next ones you sample will probably be delicious.

*Amelanchier* species
(Rosaceae)

*Amelanchier alnifolia*

Saskatoons can often be picked efficiently by using a coarse-toothed comb to rake off the berries. The twigs and leaves can be removed later by floating the berries in water and skimming the debris off the top, or by the Indian method of rolling the berries down a damp board. The leaves and twigs tend to stick to the board, while the berries roll down into a container below.

Fresh saskatoon berries are delicious in pancakes, muffins, pies, jams, or jellies. As they resemble blueberries in appearance and taste, they can be substituted for blueberries or huckleberries in cooking and baking. Canning or freezing saskatoons is highly recommended; they make an excellent breakfast fruit and can be used year-round. They are easily dried in the sun or in a food dehydrator, and can be used like currants in puddings and muffins or eaten as a snack like raisins. In short, saskatoons are an ideal wild fruit. In some areas of Canada, notably the British Columbia interior and the Prairies, saskatoons are the most abundant and important of all wild fruits.

## Suggested Recipes

### Saskatoon–Cranberry Dessert

| | | |
|---|---|---|
| 500 mL | fresh saskatoon berries | 2 cups |
| 250 mL | cranberry sauce | 1 cup |
| 250 mL | sugar | 1 cup |
| 1 mL | salt | $1/4$ tsp |
| 250 mL | marshmallows, miniature or quartered | 1 cup |

Place all ingredients except marshmallows in a blender. Cover and blend until smooth. Pour into glasses, top with marshmallows, and refrigerate 1 hour or more before serving. Serves 4–6.

## Saskatoon Pie

|  | pastry for 2-crust pie |  |
| --- | --- | --- |
| 1 L | saskatoon berries | 4 cups |
| 50 mL | sifted flour | 1/4 cup |
| 125 mL | sugar | 1/2 cup |
| 5 mL | salt | 1 tsp |
| 50 mL | butter *or* margarine | 1/4 cup |
| 30 mL | lemon juice | 2 tbsp |

Line a medium-sized pie plate with pastry, and fill with washed, drained berries. Mix sugar, flour, and salt, and sprinkle over the berries. Dot with butter, sprinkle lemon juice over, and cover with top crust. Perforate the crust with a fork and carefully seal the edges so that the juice will not overflow in the oven. Bake 45 minutes at 200°C (400°F).

## Saskatoon Surprise

| 500 mL | fresh *or* frozen saskatoon berries | 2 cups |
| --- | --- | --- |
| 250 mL | orange juice | 1 cup |
| 125 mL | vanilla ice cream | 1/2 cup |
| 250 mL | finely cracked ice | 1 cup |

Place all ingredients in a blender. Cover and blend until the berries are liquefied and the ice is melted. Serve in frosted glasses on a hot summer day. Serves 2–4.

## Pickled Saskatoons

| | | |
|---|---|---|
| 750 mL | sugar | 3 cups |
| 250 mL | cider vinegar | 1 cup |
| 5 mL | whole cloves | 1 tsp |
| 2.5 kg | saskatoon berries | 5 lb |

Boil the sugar, vinegar, and cloves together for 10 to 15 minutes. Add the berries, well washed and drained, bring to a boil, simmer for 10 minutes, and seal in sterilized jars. Makes about 6 medium-sized jars.

## Spiced Saskatoons

| | | |
|---|---|---|
| 500 mL | sugar | 2 cups |
| 250 mL | water | 1 cup |
| | 1 lemon, sliced | |
| | 4 whole cloves | |
| 2 mL | almond extract | 1/2 tsp |
| 3 L | fresh saskatoon berries | 12 cups |

Make a syrup of the sugar and water and bring to a boil. Add the remaining ingredients, bring to a boil again, and boil for 8 to 10 minutes, stirring frequently. Pour into hot, sterilized jars and seal with melted paraffin. Store in a cool place. Makes 6 medium-sized jars.

## Saskatoon Pemmican

| | | |
|---|---|---|
| 250 mL | dried beef *or* venison (jerky) | 1 cup |
| 250 mL | dried saskatoon berries | 1 cup |
| 250 mL | unroasted sunflower seeds *or* nuts of any kind | 1 cup |
| 10 mL | honey | 2 tsp |
| 50 mL | peanut butter | 1/4 cup |
| 2 mL | cayenne pepper (optional) | 1/2 tsp |

Pound or grind the dried meat to a powder. Add the dried saskatoons and sunflower seeds *or* nuts. Heat together honey, peanut butter, and cayenne pepper until soft. Blend into the other ingredients. Place in plastic bags or sausage casings (available at many butcher shops). Store in a cool, dry place where it will keep for many months. The same recipe can be made using dried blueberries or other wild fruits. Makes about 750 mL (3 cups) pemmican.

## More for Your Interest

In comparison with other dried and fresh fruits, saskatoon berries contain unusually high concentrations of iron and copper. The amount of iron is three times that contained in dried prunes and four times that in raisins.

Saskatoons have been used for centuries by native peoples in North America, and in many cases have been a staple food, more important than any other wild fruit or vegetable. To be dried for winter use they were spread out on mats, or after mashing and boiling were set out to dry in cakes. Among the Plains Indians, saskatoons were a major constituent of pemmican, which was used during the winter or when travelling (see recipe for Saskatoon Pemmican). The berries were also cooked with various bulbs and roots and with venison or bear grease in soups and stews, or were mixed with other less palatable fruits as a sweetener.

Saskatoon wood was once used by Indian peoples of British Columbia to make arrows. The long, straight sucker shoots were first chewed to break the grain and prevent warping, then hardened by baking in the fire.

The origin of the name *saskatoon* is described in the 1977 municipal manual of the City of Saskatoon (p. 1):

The name 'Saskatoon' is derived from the Cree Indian name 'Mis-sask-quah-toomina'. These words are the plural. The singular is secured by dropping the final 'a'. The name was given by the Indians to the berry [*A. alnifolia*] which is found in such profusion in this vicinity. According to a member of the original townsite survey party, the Indians used to gather large quantities of these berries and walk through the camps selling them crying what sounded to the members of the party 'Saskatoons, Saskatoons'.

# Wild Hawthorns

(Rose Family)

**Other Names**
Thornberry, haw.

**How to Recognize**
There are over one hundred different species of hawthorn native to North America, twenty-five or thirty of which can be found in various parts of Canada, most of them in the East. The hawthorns are generally similar in appearance, being small trees or large shrubs, under 8 to 10 m tall, with rough, shreddy bark, conspicuous spines or thorns on their branches, and simple, alternate, coarsely toothed or sometimes lobed leaves, which are deciduous. The flowers are 5-petalled, white or pinkish, and arranged in flat-topped clusters at the ends of short twigs. They are often very showy. The fruits, or haws, resemble miniature apples, but are clustered, thick-skinned, and range in colour from reddish orange to scarlet to black. The fruits of all species are edible, though some are better-flavoured than others. Most contain 3 to 5 large seeds.

Notable native hawthorns in Canada include the succulent-fruited hawthorn

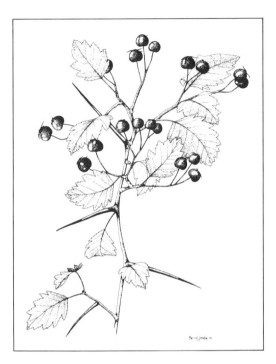

***Crataegus* species**
(Rosaceae)

*Crataegus succulenta*

(*C. succulenta* Link), a shrub or tree up to 8 m high, with usually thorny branches, sharply serrated leaves, and bright-red succulent fruit; the golden-fruited hawthorn (*C. chrysocarpa* Ashe), a large shrub with slender thorns, almost circular shallowly lobed leaves, and reddish-orange fruit; the long-spined, red-fruited Columbia hawthorn (*C. columbiana* Howell); the stout-thorned, black-fruited black hawthorn (*C. douglasii* Lindl.); the dotted hawthorn (*C. punctata* Jacq.), with conspicuously spotted fruit; the showy, large-flowered Quebec hawthorn (*C. submollis* Sarg.); the fuzzy-leaved downy hawthorn (*C. mollis* Scheele), which has particularly large fruits; and the thick-spined, scarlet-fruited cockspur hawthorn (*C. crusgalli* L.). There are many others and new forms and varieties are constantly appearing, as the hawthorns tend to hybridize freely.

In addition to the native hawthorn species, two European species introduced as ornamentals are often found as escapes in parts of southern Canada. They are the maytree, or one-seeded hawthorn (*C. monogyna* Jacq.), and the English hawthorn (*C. oxyacantha* L.). Both have short thorns, deeply lobed leaves, and red fruits, which are more astringent than native haws.

**Where to Find**
Hawthorns are found throughout the southern half of Canada in woods ranging from dry to moist, in thickets and clearings, and along roadsides. The succulent-fruited hawthorn is found from southern Manitoba east to Nova Scotia and Prince Edward Island. The golden-fruited hawthorn ranges from Newfoundland to the Rocky Mountains. Columbia hawthorn occurs, in Canada, only in south-central British Columbia. Black hawthorn is found throughout most of British Columbia, and ranges east to Saskatchewan and some localities near Lake Superior. The remaining species occur in the southeastern parts of the country, mainly southern Ontario and Quebec.

## How to Use

Hawthorn fruits have not been very popular for eating because they are somewhat dry and mealy and contain large seeds. However, in many species they are sweet and good-flavoured and excellent for making jelly. They can usually be gathered in quantity because they grow in large clusters. They ripen in late summer and early fall, but often stay on the bushes into winter. Children may like to nibble them raw, but the seeds are a nuisance, and the fruits of some species are said to cause stomach upset if too many are consumed. The fruits can be mixed with other types in making jelly, and the pulp can be made into sauce or used in various beverages. Keep in mind that some species make better jelly than others; in general, the larger and sweeter-tasting the fruits, the better the jelly will be.

## Warning

Hawthorn berries can be toxic if consumed in large quantities. The berries and leaves have long been used medicinally for their cardiotonic effect on the heart. They contain a glycoside with digitalis-like properties.

## Suggested Recipes

### Hawthorn Jelly

| | | |
|---|---|---|
| 1 kg | hawthorn fruits | 2 lb |
| 500 mL | water | 2 cups |
| 5 mL | lemon juice | 1 tsp |
| | sugar | |

Wash and de-stem the hawthorn fruits and place them in a large saucepan with the water. Heat and simmer until the fruits are soft. Mash well and allow to drain through a jelly bag for several hours, or until dripping stops. Measure juice and return to pan. Add lemon juice and the same volume of sugar as hawthorn juice, stir over low heat until sugar is dissolved, then boil until the jelly sets when tested on a cold plate. Pour into hot, sterilized jars and seal with melted paraffin. This jelly is delicious with cold meats, hot rolls, or toast. Makes about 4 large jelly glasses.

## Haw and Damson Sauce

| | | | |
|---|---|---|---|
| 500 g | hawthorn fruits | 1 | lb |
| 250 g | damson plums | ½ | lb |
| 1 L | water | 1 | qt |
| 25 mL | whole allspice | 1½ | tbsp |
| 25 mL | cloves | 1½ | tbsp |
| 5 mL | salt | 1 | tsp |
| 2 mL | cayenne | ½ | tsp |
| 500 g | sugar | 1 | lb |
| 1 L | vinegar | 1 | qt |

Wash and de-stem haws and damsons and boil in water until soft. Rub through a sieve to remove seeds and return to pan. Add allspice and cloves (tied in a cheesecloth bag) and the other ingredients. Stir until sugar is dissolved and bring to a boil. Boil until thick (about 1 hour). Remove spices, then bottle and seal, using hot, sterilized jars. An excellent sauce for ham, chicken or duck. Makes about 4 medium-sized jars.

## More for Your Interest

On the Kamchatka Peninsula, across the Bering Sea from Alaska, hawthorn fruit was fermented in water to make wine. The fruit was also popular among North American Indians; in British Columbia black hawthorn fruits were often mashed and dried into cakes for winter, and were commonly eaten with choke cherries and saskatoon berries.

North American Indians used the long, sharp spines of the hawthorns as probes, awls, and sometimes fish-hooks. Hawthorn wood is hard and tough and was used by native peoples to make digging sticks, clubs, and implement handles.

# Wild Strawberries

(Rose Family)

## How to Recognize

There are at least three species of wild strawberry in Canada: the woodland strawberry (*F. vesca* L.); the Virginia, or field, strawberry (*F. virginiana* Duchesne); and the seaside, or Pacific, strawberry [*F. chiloensis* (L.) Duchesne]. The first two are usually divided into several varieties, some of which are recognized as separate species by some botanists. Additionally, the garden strawberry, considered to be a hybrid (sometimes designated *F. chiloensis* var. *ananassa* Bailey) of the seaside strawberry and Virginia strawberry, is commonly grown throughout Canada, and is becoming a frequent escape along roadsides and railways. The fruits of all species are delicious and can be used interchangeably in cooking.

Almost everyone can recognize a wild strawberry because the fruits are so well known and the plants so similar to those under cultivation. The wild strawberries, like their cultivated relatives, are leafy perennials growing from thick rootstocks. They reproduce vegetatively by means of long runners, or stolons. The leaves have long stalks and are divided into 3 more or less equal rounded segments, each coarsely toothed at the edges. The flowers, in loose,

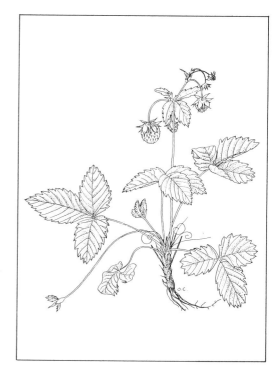

***Fragaria* species**
(Rosaceae)

*Fragaria vesca*

few-flowered clusters, are white and 5-petalled. The fruits, which develop from the flower receptacles, are spherical or conical, fleshy, and usually bright scarlet, with tiny, dry "seeds" (achenes) adhering to the outer surface. The woodland strawberry has bright-green or yellowish-green leaves, with the terminal tooth on the leaflets quite prominent. The fruits tend to be long-stemmed and elongated, and the achenes adhere superficially to the fruit. In the Virginia and seaside strawberries the leaves are more bluish green, with the terminal tooth on the leaflets shorter than those on either side. The fruits are often produced at ground level, or at least below the level of the leaves, and are more spherical, with the achenes set in pits on the fruit. The seaside strawberry has thicker, more leathery leaves than the Virginia strawberry, and is strictly coastal in its distribution.

## Where to Find

The woodland and Virginia strawberries are found in woods, meadows, and clearings across Canada, from lowlands to consider-able elevations in the mountains. The seaside strawberry is found only along the Pacific coast, on sand dunes and rocky headlands.

## How to Use

There are few wild fruits anywhere that can compare in flavour with a succulent, ripe wild strawberry. Although small in size, wild strawberries often occur in large numbers, especially in moist clearings. They usually ripen in June (at lower elevations) and are thus among the earliest wild fruits. They are best when eaten fresh, seldom require sweetening, and are so fragrant that before eating them you should close your eyes and breathe in their sweet aroma, which is equal to the best perfume money can buy. It is said that in some places you can smell a patch of wild strawberries long before you come across them.

Wild strawberries can be used in any way that the cultivated ones are. They contain less water than garden strawberries and can be easily dried if you can gather enough. Whatever they lack in size they certainly make up in flavour. In short, we concur with fisherman-philosopher-writer Isaac Walton, who said that doubtless God could have made a better berry, but doubtless He never did.

## Suggested Recipes

### Wild Strawberry Shortcake

| | | | |
|---|---|---:|---|
| 500 mL | all-purpose flour | 2 | cups |
| 20 mL | baking powder | 4 | tsp |
| 2 mL | salt | $^1/_2$ | tsp |
| 50 mL | sugar | $^1/_4$ | cup |
| 50 mL | shortening *or* butter | $^1/_4$ | cup |
| 200 mL | milk | $^3/_4$ | cup |
| 125 mL | whipped cream | $^1/_2$ | cup |
| 500 mL | wild strawberries, mashed slightly | 2 | cups |
| | whole wild strawberries for garnish | | |

To make the sweet biscuits for shortcake, mix and sift dry ingredients in a large bowl. Cut in shortening *or* butter until the largest bits of fat are smaller than peas. Slowly add milk, mixing with long, quick strokes, until the dough is soft but not sticky and comes freely from the side of the bowl. (You may need less than this amount of milk). Turn dough out on a sparsely floured board and knead very lightly for a few seconds. Roll out gently with a rolling pin or pat out to about 2 cm ($^3/_4$ in.) thick. Cut into circles about 6 cm ($2^1/_2$ in.) across with floured cutter and place on greased cookie sheet. Bake in preheated oven at 220°C (425°F) for 12 to 15 minutes, or until biscuits are well risen and lightly browned on top.

Meanwhile, prepare strawberries, and whip cream until stiff. While biscuits are still warm, split four in half (one per serving) and place each in a dish. Spoon about 50 ml ($^1/_4$ cup) strawberries over the bottom half, cover with the top half, and add more strawberries, allowing 125 ml ($^1/_2$ cup) mashed strawberries per person. Top with whipped cream (sweetened if desired) and sprinkle a few whole wild strawberries on top as a garnish. Serve immediately. This entire dessert can be made in less than half an hour if the berries are prepared beforehand. Serves 4, with several biscuits left over. Note: half the amount of white flour may be replaced with whole-wheat flour for an interesting variation.

## Wild Strawberry Delight

| | | | |
|---|---|---|---|
| 125 mL | fruit sugar | 1/2 | cup |
| 125 mL | whipped cream | 1/2 | cup |
| 250 mL | mashed wild strawberries | 1 | cup |
| 25 mL | brandy *or* Cointreau | 1 1/2 | tbsp |
| | a small handful of fresh strawberries | | |

Beat egg whites until stiff, and fold in sugar. Whip cream. Combine egg whites, whipped cream, mashed strawberries, and brandy *or* Cointreau. Sprinkle with whole berries and serve immediately. Serves 2–3. (From Enid K. Lemon and Lissa Calvert, "Pick'n' Cook Notes".)

## Strawberry "Leather"

Gather as many ripe wild strawberries as you can. Mash by hand or purée in a blender, and pour out onto large sheets of heavy waxed paper. Place these in the sun or in a food dehydrator, and allow the fruit to dry to a tough, leather-like consistency. Peel off the paper and store the "leather" in jars in a cool place. Will keep well if thoroughly dried.

## Sun-cooked Strawberry Jam

| | | |
|---|---|---|
| 2 L | wild strawberries | 8 cups |
| 1.3 L | sugar | 5 1/2 cups |

Carefully wash, drain, and hull berries. Combine fruit with sugar in alternate layers in large saucepan. Bring to boiling point slowly over low heat, stirring until sugar is dissolved. Boil rapidly for 2 minutes. Skim. Spread the berry mixture in thin layers on large platters or baking pans. Cover with window or picture glass and let stand in the sun for 7 to 10 hours. Pour into hot, sterilized jars and seal with paraffin. Store in a cool place. Yields about 3 small jars. Much of the fresh strawberry flavour is retained using this method.

## More for Your Interest

The name *strawberry* is not derived, as many people think, from the berries being grown over straw to keep them clean, but from the Anglo-Saxon name *streowberie*, alluding to the plant's long runners being "strewed" over the ground.

Strawberry leaves make a very pleasant tea, used for centuries as a folk remedy for diarrhoea. The crushed berries also make a pleasant beverage, alone or mixed with other fruit juices in punch. (See our publication *Wild Coffee and Tea Substitutes of Canada*.)

# Canada Plum and American Plum

(Rose Family)

### Other Names
Both species are called wild plum.

### How to Recognize
These are both rather crooked, straggling trees with stiff, spreading branches forming an irregular crown; they reach a height of about 8 m. The bark of Canada plum is black, while that of American plum is usually more reddish brown or dark grey. Both have short horizontal markings, or lenticels, on the young bark. The leaves of both species are oval, but those of Canada plum are wider and usually broadest above the middle of the blade, whereas those of American plum are narrower and usually broadest below the middle, tapering gradually to a sharp tip. The blossoms appear with or before the leaves in spring. They are large, numerous, and showy, with white petals, which are often tinged with pink in the Canada plum. They are usually borne in loose clusters of 3 to 5 flowers. The fruits are reddish orange, about 2.5 cm long, with a slight waxy coating, or bloom, in American plum, and no bloom in Canada plum. They ripen in late summer and are juicy and sweet when fully mature, although the skins tend to be puckery and the flesh around the pits sour. The pits, like those of other plums, are flattened. This is one of the main distinguishing features between plums and their close relatives, the cherries, which have small, nearly spherical pits.

Canada plum and American plum are really very similar, and some botanists include them as varieties of the same species. They are the only native plums in Canada, but many types have been introduced from other countries and some, particularly the garden plum (*P. domestica* L.) and the sloe plum (*P. spinosa* L.), are occasionally found as garden escapes in various parts of Canada.

*Prunus nigra* **Ait. and** *P. americana* **Marsh.**
(Rosaceae)

*Prunus americana*

## Where to Find

Canada plum is a scattered but widespread tree of river valleys and limestone soils from southern Manitoba through southern Ontario and Quebec to New Brunswick. American plum is native only to southern Ontario, southern Manitoba, and the extreme south-eastern corner of Saskatchewan, but has been widely planted as an ornamental and fruit tree and is now often found growing wild well beyond its natural range. It grows in rich soils along streams and marsh borders.

## How to Use

Before they are completely ripe, these plums are tough-skinned and astringent, but once ripened they are sweet and soft to the touch, with deep, translucent yellow flesh. You will find no commercial plum to compare with them.

The best way to harvest wild plums is to shake the trees, at first lightly to dislodge the ripest fruits, which should be gathered separately, and then more vigorously to get those slightly less ripe. Take care not to damage the tree, however, as the branches are quite brittle.

Wild plums make a superb jelly, alone or mixed with other fruits. They can also be used to make jams, sauces, and fruit juices, and can be dried and used as raisins, or puréed, poured out on waxed paper, and dried as fruit "leather". To make plum jams and jellies use about half ripe and half unripe fruits and you will not need to use commercial pectin. Wild plums are also excellent in pies, and can be substituted for domesticated plums or cherries in most baking recipes.

## Warning

The Warning on p. 153 for pin cherry (*Prunus pensylvanica*) is applicable to wild plums as well.

## Suggested Recipes

### Wild Plum Muffins*

| | | | |
|---|---|---|---|
| 425 mL | flour | 1³/₄ | cups |
| 15 mL | baking powder | 1 | tbsp |
| 2 mL | salt | ¹/₂ | tsp |
| 50 mL | wheat germ | ¹/₄ | cup |
| 50 mL | brown sugar | ¹/₄ | cup |
| 250 mL | wild plums, with skins on, cut in small pieces | 1 | cup |
| | 1 egg, beaten | | |
| 250 mL | milk | 1 | cup |
| 30 mL | vegetable oil | 2 | tbsp |

Sift flour, baking powder, and salt together. Add wheat germ and sugar. Add well-drained, chopped plums and toss until coated with flour. Mix liquid ingredients together and fold into flour mixture. Spoon into large buttered muffin tins, and bake at 180°C (350°F) for 25 minutes. Yields about 1¹/₂ dozen medium-sized muffins.

### Lamb Shanks with Wild Plums*

| | | | |
|---|---|---|---|
| | 4 lamb shanks | | |
| 50 mL | flour | ¹/₄ | cup |
| 500 mL | quartered, pitted wild plums | 2 | cups |
| 50 mL | sugar | ¹/₄ | cup |
| 250 mL | water | 1 | cup |
| 50 mL | apple vinegar | ¹/₄ | cup |
| 2 mL | cinnamon | ¹/₂ | tsp |
| 2 mL | ground cloves | ¹/₂ | tsp |
| 2 mL | allspice | ¹/₂ | tsp |
| 2 mL | salt | ¹/₂ | tsp |
| 15 mL | cornstarch (optional) | 1 | tbsp |

Season the meat with salt and pepper, and dredge with flour. Place in a greased baking dish, cover, and bake at 180°C (350°F) about 2 hours.

Combine all the remaining ingredients and simmer 5 minutes to blend. Skim fat from the baking dish and add the fruit sauce to the meat, then cover and bake for an additional 30 minutes. Place the shanks in a serving dish. Strain the sauce, thicken with a little cornstarch if you wish, and pour over the meat. Serve very hot.

*These recipes were modified from Constance Conrader's in "Wild Harvest".

## Wild Plum Ice Cream*

| | | | |
|---|---|---|---|
| 125 mL | sugar | ½ | cup |
| 50 mL | water | ¼ | cup |
| 2 mL | cream of tartar | ½ | tsp |
| | 4 egg yolks | | |
| 30 mL | crystallized ginger, cut very fine | 2 | tbsp |
| 750 mL | cream | 3 | cups |
| 250 mL | fresh wild plum nectar | 1 | cup |

Boil sugar, water, and cream of tartar to 110°C (230°F), or the thread stage. Beat egg yolks, then slowly pour hot syrup over them, beating until they reach the consistency of thin whipping cream. Add the finely cut ginger and the cream. Turn into an ice-cream freezer and proceed according to your freezer directions. When ice cream is partially frozen, add the wild plum nectar (made by simmering, mashing, and straining wild plums) and freeze at least 2 hours before using.

## More for Your Interest

Both Canada plum and American plum are often planted in gardens, and several cultivated forms have been developed by horticulturalists.

According to R.C. Hosie, in *Native Trees of Canada*, potatoes should not be planted near plum trees because of an aphid that sometimes overwinters in the tree. Although it is relatively harmless to the trees and their fruit, the aphid can carry a potato disease.

## Other Names

Red cherry, bird cherry, fire cherry, Pennsylvania cherry.

## How to Recognize

There are five species of cherry native to Canada. Although closely related to the plums, the cherries have smaller fruits with almost spherical pits. The plums have larger, more fleshy fruits and more flattened pits. The most important and widespread wild cherries are pin cherry and choke cherry (*P. virginiana*). The latter is discussed, along with wild black cherry (*P. serotina*), in the next section.

Pin cherry is an attractive, straight-trunked tree or tall shrub, considered by some to be the most beautiful of all the cherries. It can reach heights of 10 to 12 m but is usually smaller. The bark on young trees is smooth, with a dark reddish-brown, varnished appearance. The bark on mature trees is marked with widely spaced horizontal lines, or lenticels. The leaves are thin, fragile, and lance-shaped, being widest below the middle of the blade and tapering gradually to a slender, sharp tip. The edges are finely toothed and the midrib is prominent. The white-petalled, showy flowers appear with the leaves in spring. The cherries are small and bright red, with tart but edible flesh and rather large stones. They ripen in late summer.

The most obvious difference between pin cherry and choke cherry is in the nature of the flowering and fruiting clusters. In pin cherry the flowers and fruits are borne in loose clusters of 5 to 7, each fruit with a long, slender stalk attaching it to a common point on the twig. Choke cherry flowers and fruits, and those of wild black cherry, are borne in elongated cylindrical clusters, the individual flowers and fruits on short stems spaced along a central axis.

Two other species of wild cherry should be mentioned. One, the sand cherry (*P. pumila* L.), is a low shrub that grows on sandy soil. The fruits are dark purple, in clusters of 2 to 3, and when ripe are pleasantly acid, sometimes slightly bitter. The sand cherry is highly variable and is separated by some botanists into several species. The other notable cherry, known as bitter cherry [*P. emarginata* (Dougl.) Walpers], is a tree similar in size and appearance to pin cherry but with oval leaves,

# Pin Cherry

(Rose Family)

tapering at both ends and more rounded at the tip than pin cherry leaves. The fruit looks inviting, but in most cases is extremely bitter and astringent. However, occasional populations of bitter cherry can be found with quite edible cherries. None are poisonous, and you need only taste one fruit to know whether they are worth harvesting.

## Where to Find
Pin cherry grows in wooded areas and along river banks from Newfoundland to central British Columbia, extending north as far as Great Slave Lake in the Northwest Territories. The sand cherry grows on dunes, beaches, sandy barrens, and calcareous outcrops from Manitoba to New Brunswick. Bitter cherry is common in woods and thickets in the southern half of British Columbia.

## How to Use
It is sometimes difficult to beat the birds to the valuable pin cherry crop. If you are fortunate enough to harvest a good quantity of these small, delightfully tart fruits, they make one of the best jellies to be had, and are also excellent for making juice, syrup, wine, and sauce. Removing the pits takes considerable time and patience, but pin cherry pie and pin cherry ice cream are delicious, and of course cherries are good in all types of baked products—cakes, muffins, fruit breads, cobblers, biscuits, and cookies.

Pin cherries can be dried, with or without the pits, and can also be frozen or canned, although the small size of the fruits is a definite disadvantage in using them.

## Warning
The partially wilted leaves and the pits and bark of the pin cherry, like those of other *Prunus* species, liberate hydrocyanic acid (prussic acid) into the stomach. The leaves of wild *Prunus* species have been a frequent cause of livestock poisoning. Used as flavouring and in folk remedies, the leaves and bark have caused many deaths. Although the flesh of the fruits is not harmful, poisoning and death have occurred in children who consumed large quantities of the fruits without removing the stones, or who have chewed the twigs.

**Suggested Recipes**

## Pin Cherry Jelly

| | | | |
|---|---|---|---|
| 1.5 L | pin cherries | 6 | cups |
| 250 mL | water | 1 | cup |
| 57 g | powdered pectin (1 package) | 2 | oz |
| 1.2 L | sugar | 4½ | cups |

Wash cherries and place in a large saucepan with the water. Bring to a boil, reduce heat, and simmer for 30 minutes, mashing slightly to press out the juice. Strain through a jelly bag overnight or until dripping stops. Return 875 mL (3½ cups) juice to saucepan and mix in pectin crystals. Place over high heat and stir until mixture comes to a hard boil. Stir in sugar and return to a full, rolling boil. Boil hard 1 minute, stirring constantly, then remove from heat, skim off foam, and pour into hot, sterilized glasses. Seal with melted paraffin and store in a cool place. Makes about 8 medium-sized jelly glasses. This jelly is bright red and pleasantly tart to taste.

## *Sauce Cerise* for Roast Chicken or Other Fowl

| | | | |
|---|---|---|---|
| 500 mL | chicken stock | 2 | cups |
| 175 mL | pin cherry jelly | ¾ | cup |
| | 2 oranges | | |
| | 1 lemon | | |
| 50 mL | claret wine | ¼ | cup |
| | dripping from roast chicken | | |

Simmer chicken stock and jelly together on low heat for 1 hour. Meanwhile, grate and squeeze the oranges and lemon and simmer the juice with the rind essences for 15 minutes. Add to the stock, together with the wine, and simmer very gently (don't boil) for 10 minutes more. Strain. Just before serving, reheat to steaming and add the strained fat-free dripping from the roasting pan. Carve the chicken and dip pieces into the sauce as they are served. (From Constance Conrader, "Wild Harvest".)

## Wild Cherry Pudding

| | | |
|---|---|---|
| 250 mL | wild cherries, pitted | 1 cup |
| 250 mL | brown sugar | 1 cup |
| 50 mL | butter | 1/4 cup |
| 125 mL | white flour | 1/2 cup |
| 125 mL | whole-wheat flour | 1/2 cup |
| 5 mL | baking powder | 1 cup |
| | pinch of salt | |
| 2 mL | allspice | 1/2 tsp |
| 2 mL | mace | 1/2 tsp |
| 125 mL | milk | 1/2 cup |

Combine cherries with one-half of the sugar and heat. Cream together butter and remaining sugar, sift together dry ingredients and spices, and add alternately with milk to butter mixture. Pour into a greased baking dish. Pour hot cherries over the batter and bake at 150°C (300°F) for 45 minutes. Serves 4.

## More for Your Interest

The bark of pin cherry and other wild cherries has long been used by both Indians and white settlers to make a tea for coughs and stomach ailments, although this is not recommended (see Warning above). A syrup made from the fruits is also said to be good for coughs, and has been used in the past as a pleasant syrup in which to dilute medicines.

The smooth, tough bark of both pin cherry and its relative, the bitter cherry, pulls off the trunk in long strips, in the same manner as birch bark. Among the Indians of British Columbia, bitter-cherry bark was an important material for wrapping and strengthening the joints of implements such as harpoons, spears, and bows and arrows. When glued in place with tree pitch and bound with sinew or plant fibre, it made a neat, watertight bond. It was also used in its natural reddish colour or dyed black to superimpose designs on coiled-root baskets. Pin cherry bark may have been used similarly in other parts of Canada.

# Choke Cherry

(Rose Family)

## Other Name
Wild cherry.

## How to Recognize
This is a low or tall shrub or small tree up to 8 m in height, with mostly upright, often crooked, slender branches and smooth or scaly, dark-greyish bark without noticeable horizontal marks. The leaves are simple, broadly oval, abruptly pointed at the tips, and finely toothed around the margins. They vary from 8 to 10 cm in length. The flowers are small and whitish or creamy, in dense, long, bottlebrush-like clusters, appearing after the first leaves are almost fully developed. They have a strong, rather unpleasant odour. The cherries are pea-sized and fleshy but with large stones. There are different varieties, ranging in colour from bright, translucent red to deep reddish purple or black. The fruits are quite sweet and palatable when fully ripe, but astringent and "choky" when unripe—hence the name *choke cherry.*

A related species, the black cherry (*P. serotina* Ehrh.), also has fruits in long clusters, but is usually a very tall tree, up to 30 m or more in height, with narrowly oval, tapering leaves and a conspicuous 5-pointed calyx at the base of each fruit. The cherries are slightly bitter but quite edible.

## Where to Find
Choke cherries occur across Canada from the Pacific coast to Newfoundland and as far north as the southern Yukon. There are different varieties but all are recognizable as choke cherries. They are commonly found in open situations, on rich, moist soils—along fencelines, streams, and the edges of woodlands—and sometimes on rocky slopes. Black cherry is restricted to the deciduous forest region of the Great Lakes and St. Lawrence areas and the Maritimes.

## How to Use
Choke cherries are easy to harvest in large quantities because they grow in such long, dense clusters. Unfortunately, the large seeds make them difficult to use in baking, but if you have the patience to extract the pits, the cherries make delicious pies and other desserts. Choke cherries are also excellent for juice, jelly, and syrup, and blend well with other fruit juices. If you eat the fruits before they are completely ripe they leave a "cottony" sensation in your mouth, and you won't want to eat very many. You should avoid swallowing the seeds as too many will make you sick (see Warning below).

Choke cherries can be dried and also freeze well; they are said to lose some of their astringency after being frozen. Black cherries can be used in the same way as choke cherries.

*Prunus virginiana* L.
(Rosaceae)

**Warning**

The warning on p. 153 for pin cherry (*Prunus pensylvanica*) is applicable to choke cherries as well.

**Suggested Recipes**

### Choke Cherry Wine

| | | | |
|---|---|---|---|
| 9 L | ripe choke cherries | 8 | qt |
| 14 L | water | 12 | qt |
| | sugar | | |
| 8 g | dry yeast (1 package) | 1/4 | oz |
| 125 g | chopped raisins (optional) | 1/4 | lb |
| 250 mL | good brandy (optional) | 1 | cup |

Place cherries and water in large saucepan, bring to a boil, mash, and simmer 10 minutes. Strain off juice through a fine sieve or jelly bag. Measure and place in large crock or wine keg, add 750 g (1 1/2 lb) sugar per 1 L (4 cups) juice, mix, and cool to lukewarm. Add yeast and raisins (optional), mix well, cover, and ferment for 5 days or until bubbling has ceased. Stir well and add brandy if desired. Keep covered for about 3 months, then filter and bottle. After a few days, cork tightly and store in a dark, cool place for 6 months before using.

### Choke Cherry–Apple Jelly

| | | | |
|---|---|---|---|
| 1 L | ripe choke cherries | 4 | cups |
| 125 mL | water | 1/2 | cup |
| 875 mL | unsweetened apple juice | 3 1/2 | cups |
| 30 mL | lemon juice | 2 | tbsp |
| 57 g | pectin crystals (1 package) | 2 | oz |
| 1.5 L | sugar | 3 | lb |

Stem but do not pit the choke cherries, and place in a large saucepan with the water. Mash and simmer, covered, for 10 minutes. Strain through a jelly bag and measure juice. Place 425 mL (1 3/4 cups) juice back in saucepan, add apple juice and lemon juice, and mix in pectin crystals. Place over high heat, stirring constantly, until mixture comes to a full boil. Stir in sugar. Bring to a boil again and boil hard 1 minute, stirring continuously. Then remove from heat, skim, and pour into hot, sterilized jars. Seal with melted paraffin. Store in a cool place. Makes about 10 to 12 medium-sized jelly glasses.

## Choke Cherry Syrup

| 2 L | ripe choke cherries | 8 cups |
|---|---|---|
| 125 mL | water | ½ cup |
| 1 kg | sugar | 2 lb |
| 30 g | pectin crystals | 1 oz |

Make choke cherry juice as for jelly in previous recipe. Put juice (about 1 L) in saucepan, add pectin crystals, and bring to a boil, stirring frequently. Stir in sugar, boil hard 1 minute, then skim and pour into hot, sterilized jars. Store in refrigerator. Excellent on pancakes. Makes about 3 medium-sized jars.

## More for Your Interest

Choke cherries were among the most important of Indian fruits, being second only to saskatoon berries in some areas of western Canada. They were often preserved by drying, and were a common ingredient of pemmican, made with dried meat, melted animal fat, and dried fruits (see recipe, p. 136). In the British Columbia interior, dried choke cherries were often eaten with salmon or salmon eggs. The Okanagan Indians also used the dried cherries for making a tea that was considered especially good for coughs and colds. The wood, branches, and bark were also used to make a tea for the treatment of colds and diarrhoea (but see Warning, p. 153). The Shuswap Indians of British Columbia mixed the fruits with bear grease to make a paint for colouring pictographs. Choke cherry wood was sometimes used by Indian people to make digging sticks, roasting skewers, and fire tongs.

# Wild Crabapples

(Rose Family)

### Other Names
*P. coronaria* is sometimes called sweet or eastern crabapple; *P. fusca* is known as Pacific, western, or Oregon crabapple.

### How to Recognize
Apples are closely related to pears and also to the mountain-ashes. Some botanists place all three in the same genus, *Pyrus*, although the mountain-ashes are usually separated into the genus *Sorbus*. Sometimes apples are also placed in a separate genus, *Malus*, and pears are retained in the genus *Pyrus*. The main difference between apples and pears is in the fruit: pears have hard "stone" cells embedded in the flesh and a distinctive "pear" shape, whereas apples have no stone cells and have a more spherical or oval shape. Only two *Pyrus* species—those listed here—are native to Canada, but cultivated apples and pears have become naturalized in many parts of the country, especially along roadsides.

The native crabapples are deciduous shrubs or small trees up to 12 m in height and 40 cm in diameter. The twigs often bear long thorns or thorn-like spurs. The sweet crabapple has a rather short trunk, with

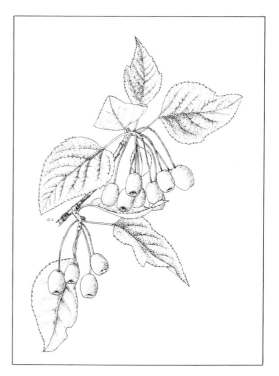

*Pyrus coronaria* L. and *P. fusca* Raf.
(Rosaceae)

*Pyrus fusca*

reddish-brown, scaly bark and spreading limbs forming a round-topped crown. The leaves are irregularly toothed, often lobed several times. The flowers are rosy white, smaller than those of orchard apple trees, and arranged in flat clusters. The fruit, up to 5 cm across, is yellowish green and sour, but quite edible. The resemblance of the fruit to the cultivated apple makes the species easy to identify.

The Pacific crabapple is usually a taller tree, with rough, grey bark and leaves similar in shape to those of orchard apples, but often with a prominent, pointed lobe along one or both edges. The white to pinkish flowers are replaced by small, elongated, yellow to purplish-red apples in long-stemmed, hanging clusters. These are edible but very tart. After the first frost they turn brown and soft and become much sweeter. Another name for this apple is *Malus diversifolia* (Bong.) Roemer.

Wild apples and pears, which have escaped from cultivation, are usually large trees, up to 20 m in height, and similar in habit to the cultivated varieties. The fruits are large and single.

**Where to Find**
Crabapples generally grow in moist places, such as along stream banks and in swamps, bogs, and moist woods, as well as along roadsides and fences. In Canada sweet crabapple is confined to the southernmost part of Ontario, and Pacific crabapple is found along the entire coast of British Columbia west of the Coast and Cascade mountains.

**How to Use**
Crabapple fruit, though pleasing to the eye, has an acid taste, and can seldom be eaten fresh in quantity unless it is very ripe. The tart fruit does make a good thirst-quencher on a hike, however. Cooked, the fruit is excellent for various preserves. Crabapple jelly is especially delicious, and as the fruits are high in pectin, they are good in combination with wild blackberries, rose hips, and many other wild fruits that are low in pectin and would not otherwise set without commercial pectin. All crabapple preserves have a special aromatic flavour,

especially after the fruit has been touched with frost. After you make jelly save the residual pulp, first rubbing it through a sieve to remove the seeds and stems. You will find this purée has a rich, almond-like flavour from being cooked with the seeds (but see Warning below). When mixed with sugar and fermented it makes a tangy cider. It is also good in breads, cakes, and cookies. Crabapples are suitable for freezing and canning.

## Warning

Like the seeds of the *Prunus* species, those of *Pyrus* and *Malus* contain a cyanogenic glycoside that has an almond-like flavour and releases hydrocyanic acid. Small amounts for flavour are not dangerous, but toxic reactions may result from larger amounts. As with *Prunus*, the bark and leaves of apple and pear trees should not be consumed.

## Suggested Recipes

### Crabapple–Mint Jelly

| | | |
|---|---|---|
| 1 kg | crabapples | 2 lb |
| 500 mL | water | 2 cups |
| 5 mL | citric *or* tartaric acid | 1 tsp |
| | sugar | |
| | 6 sprigs of fresh mint | |

Wash crabapples and place in large saucepan with water and acid. Bring to a boil and simmer slowly until fruit is pulped and all the juice is free (about 1 hour). Strain through a fine sieve or jelly bag. Measure juice and return to the pan with 340 g ($^3/_4$ lb) sugar to each 500 mL (2 cups) juice. Add the mint, tied in a bunch. Stir until boiling and boil briskly about 10 minutes, or until jelly sets when tested on a cold plate. Remove mint, pour into hot, sterilized jars, and seal with melted paraffin. Label and store in a cool place. A few drops of green colouring may be added if desired. Yields about 6 medium-sized glasses.
Note: To make crabapple–ginger jelly, leave out the mint leaves and add 10 mL (2 tsp) of ginger to the crabapples and water *before* simmering the fruit. Omit the green colouring.

## Crabapple Juice

Wash 2 kg (4 lb) crabapples, put them in large saucepan, cover with water, and simmer for 1 hour or longer, until pulped and all the juice is free. Strain through a jelly bag or hair sieve. Return juice to pan with a few drops of lemon flavouring and sugar to taste, and simmer for 10 minutes. Bottle in hot, sterilized jars, store in a cool place, and use as a drink, diluted with water. Served hot as an "apple tea" this drink is soothing for people suffering from colds.

## Raw Crabapple Juice

Crabapple juice can be made in an electric juicer in the same manner as ordinary apple juice is made. If it is not to be drunk right away, you can prevent loss of flavour and colour by the addition of about 2 ml ($^{1}/_{2}$ tsp) of ascorbic acid to 1 L (1 qt) of the juice. Sweeten to taste with sugar or honey, and dilute with an equal amount of water if you desire.

## Crabapple Cider (British Settlers' Recipe)

| 4.5 kg | crabapples | 10 | lb |
|--------|-----------|----|----|
| 5 L | water | 4 | qt |
| 2 kg | sugar | 4 | lb |
| 125 g | raisins | 4 | oz |

Wash crabapples and pass them through a mincer. Place in a large crock, cover with the water, and leave for a week, stirring twice daily. Strain, add sugar to the liquid, and heat gently until sugar is dissolved. Add raisins. Return to crock, cover lightly, and leave to ferment for another 3 to 4 weeks. When fermentation (bubbling) has stopped, strain, bottle, and leave for a day or two, then cork securely. This cider will mature in 6 months.

## More for Your Interest

In Europe the juice extracted from raw crabapples is a well-known remedy for sprains and bruises. It is called *verjuice*, a name derived from French and alluding to its acidity. In Ireland crabapple juice is often added to cider to give it "roughness".

Pacific crabapples were an important food for all coastal Indian groups in British Columbia. Picked and stored in bunches with the stems still attached, they were eaten raw or cooked, sometimes mixed with other fruits, and usually with large quantities of eulachon grease. Boxes of crabapples were a common item of native trade and commerce. At the turn of the century a single box of crabapples preserved in water might cost about ten pairs of Hudson's Bay Company blankets (worth about ten dollars altogether). Crabapple wood, which is hard and resilient, was used by Indians to make harpoon shafts, adze handles, and digging sticks. It is excellent for small carvings and lathe work.

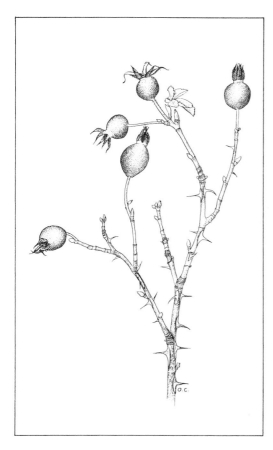

**Other Names**
Briar, bramblebush.

**How to Recognize**
There are about fourteen species of native wild roses in various parts of Canada and several more that have escaped from cultivation.

The wild roses are erect, multi-stemmed shrubs, with smooth reddish or greenish bark, usually well armed with numerous slender spines or stout thorns, or both. The leaves are pinnately divided into 5 to 9 oval-shaped leaflets with finely toothed margins. The flowers of all the native species are of the basic 5-petalled type. Some of those that have escaped from cultivation have "double" or multiple-petalled flowers. The blossoms are borne singly or in clusters, and their size varies considerably with the species. The petals, rounded or notched, are usually pale to deep pink in colour. The centres are yellowish, the stamens numerous. Most wild rose flowers are delicately and beautifully scented. Rose fruits, known as "hips", are spherical or elongated, and when ripe are orange to deep red. They consist of an edible, fleshy outer rind surrounding a mass of tightly packed whitish seeds (achenes), which are covered with many minute, sliver-like hairs. Often the pointed or leafy sepals of the flower persist in the fruit as a stiff, brownish tuft at the end of the hip.

# Wild Roses

(Rose Family)

*Rosa nutkana*

***Rosa* species**
(Rosaceae)

Some species of wild rose have hips that are too small or scarce to bother with, whereas other species have large, clustered, easily gathered hips, including the Nootka rose (*R. nutkana* Presl.), prickly rose (*R. acicularis* Lindl.), Wood's rose (*R. woodsii* Lindl.), meadow rose (*R. blanda* Ait.), and low prairie rose (*R. arkansana* Porter). Perhaps the most famous of these is the prickly rose, which is the floral emblem of Alberta.

**Where to Find**

Wild roses occur throughout Canada except above the tree line in the Far North. They grow in woods and meadows and often form dense thickets along roadsides and fences. Of the species mentioned, Nootka rose is widespread in British Columbia, prickly rose ranges from British Columbia to Quebec and northwards into the Yukon, Wood's rose is common in eastern British Columbia and the Prairie Provinces, meadow rose occurs from Manitoba to New Brunswick, and low prairie rose is found in the southwestern prairies.

**How to Use**

Rose hips are best for eating when picked fully mature but before the outer rind has been softened by frost. However, they can be gathered and used any time during the fall and winter, even when frozen on the bushes. Care should be taken to remove all the seeds, either manually or by straining the prepared juice, as the small, sliver-like hairs on the seeds can catch in the throat and digestive tract and cause irritation. Caution should also be taken when removing the seeds with bare hands, as the bristles can irritate the skin and cause dermatitis. It should be noted that the seeds do contain high concentrations of vitamin E, and, once the hairs are removed by rubbing or washing, the seeds can be ground and used in baking or cooking as a vitamin supplement.

Once the seeds are removed, rose hips can be eaten raw in salads, sandwich fillings or desserts, or can be dried for later use in soups and teas (see our previous publication *Wild Coffee and Tea Substitutes of Canada*). The raw hips are somewhat bland and uninteresting but quite acceptable when mixed with other foods.

Because rose hips are so easy to gather and so wholesome, it is surprising that they are not produced in Canada on a commercial basis. In many European countries rose hips are exported as syrup, jelly, dried soup, and tea, and most of these products are available in our own health-food stores and specialty shops, usually at a healthy price. It would seem much more economical and resourceful for us to make our own rose-hip products from our native wild roses.

## Suggested Recipes

### Rose-Hip Purée

| 1 kg | rose hips | 2 lb |
|---|---|---|
| 500 mL | boiling water | 2 cups |
| | sugar | |

De-stem and remove the hip tops and wash in lukewarm water; then place in a large saucepan and add the boiling water. Cover, bring to a boil, then reduce heat and simmer until hips are tender (about 15 minutes). Mash gently and strain through a hair sieve or cheesecloth. (If a large-mesh sieve is used, you should remove the seeds before cooking the hips.) Measure purée and mix in sugar to the proportion of 250 mL (1 cup) sugar to 500 mL (2 cups) purée. Return to heat, bring to a boil, and simmer 10 minutes. Pour into hot, sterilized jars and seal with melted paraffin. Store in refrigerator or other cool place. Yields about 8 medium-sized jars.

### Rose-Hip Syrup

| 1.5 L | rose hips | 6 cups |
|---|---|---|
| 750 mL | boiling water | 3 cups |
| 750 mL | sugar | 3 cups |

De-stem and remove the hip tops and wash in lukewarm water. Place with the boiling water in a large saucepan and boil for 20 minutes. Strain through a jelly bag overnight, or until dripping stops. Return clear juice to the pan, add the sugar, stir until dissolved, then boil 5 minutes. Pour into hot, sterilized bottles. Store in refrigerator, or cork and immerse the bottles in warm water, bring slowly to simmering point, and allow to simmer for 20 minutes. Dip tops in melted wax and cool, then store in a cool place. Makes about 1.5 L (6 cups).

## Rose-Hip Jelly

| | | |
|---|---|---|
| 1 L | rose hips | 4 cups |
| | water to cover | |
| 125 mL | lemon juice | $\frac{1}{2}$ cup |
| | sugar | |

De-stem and remove the hip tops and wash in lukewarm water; then place in a large saucepan and add enough water to cover. Add lemon juice, bring to a boil, and boil until hips are soft and mushy. Strain through a jelly bag overnight, or until dripping stops. Measure juice and return to pan with sugar to the proportion of 3 parts sugar to 4 parts juice. Stir until sugar dissolves, bring to a boil, and boil rapidly for 10 minutes, or until jelly sets when tested on a cold saucer. Pour into hot, sterilized glasses and seal with melted paraffin. Store in a cool place. Yields about 4 medium-sized glasses.

## Rose-Hip and Prune-Juice Sherbet

| | | |
|---|---|---|
| 250 mL | sugar | 1 cup |
| 250 mL | water | 1 cup |
| 125 mL | prune juice | $\frac{1}{2}$ cup |
| 125 mL | light corn syrup | $\frac{1}{2}$ cup |
| 125 mL | rose-hip syrup | $\frac{1}{2}$ cup |
| 1 mL | salt | $\frac{1}{4}$ tsp |
| | 1 egg white | |

Combine sugar and water and boil 5 minutes. Add the juice, syrups, and salt. Freeze in an ice-cube tray until almost firm. Whip egg white until stiff and fold into the frozen mixture. Return to ice-cube tray and re-freeze. Keep frozen until served. (From Eleanor A. Ellis, *Northern Cookbook.*)

## More for Your Interest

One study found that three rose hips from an Alberta population contained as much vitamin C as one orange. As a further comparison, it was found that 100 g of Alberta rose hips contained almost 1640 mg of vitamin C, about 30 times the amount contained in the same amount of pure orange juice. Rose hips are also a good source of vitamin A, calcium, phosphorus, and iron. During the Second World War, rose hips became very important in Great Britain and the Scandinavian countries when the citrus fruit supplies were cut off by German blockades. Rose hips were gathered in tremendous quantities and made into syrup, or dried and made into powder, and the product was distributed as a vitamin supplement. Rose-hip syrup is still readily available throughout the British Isles.

For centuries rose petals have been used extensively in cooking in many parts of Europe and in the Middle East. Delectable desserts such as mousses, parfaits, cakes, and candies can be made from this rather unusual food. Rose-petal nectar, rose-water, and rose-petal vinegar are also widely used. Rose-petal wine was made in England as early as 1606, and lozenges of red rose flowers were made in 1656. Rose-petal honey, made at least as long ago as the middle of the thirteenth century, is still a favourite, and easy to make. Simply take 500 mL (2 cups) of honey, bring to a boil, and add 250 mL (1 cup) wild rose petals. Let stand for a few hours, reheat, strain to remove the petals, and bottle.

The juicy young shoots of wild rose bushes can be peeled and eaten in the spring when still tender. Fruiting rose branches make excellent seasonal decorations, and the hips when threaded on a string make an interesting and unusual decoration for your Christmas tree. By the end of the Christmas season they are already dried and ready for use to make rose-hip tea.

## Other Names

*R. arcticus* is known as plumboy or arctic raspberry; *R. chamaemorus* is called cloudberry, baked-apple berry, or maltberry.

## How to Recognize

There are several species of low or creeping raspberries in Canada, the most important of which, for our purposes, are arctic raspberry (*R. arcticus*) and cloudberry (*R. chamaemorus*). Arctic raspberry is highly variable, and some botanists prefer to treat it as three separate species, including *R. arcticus* (the more typical European form), *R. acaulis* Michx., and *R. stellatus* Sm. The fruits of all are equally edible, however, and the plants hybridize freely where their ranges overlap; therefore we will treat them together here as subspecies of a single species. Other low raspberries include the hairy raspberry (*R. pubescens* Raf.) and trailing raspberry (*R. pedatus* Sm.).

Arctic raspberry is a short plant about 5 to 10 cm tall, with erect flowering shoots growing from a woody rootstock. The leaves, usually only 2 to 3 per plant, are 3-parted, or, in the case of ssp. *stellatus* (Sm.) Boiv., merely 3-lobed. The leaf margins are coarsely toothed, and the leaflets are pointed to rounded at the tips. The flowers are usually single or in clusters of 2 or 3, and the petal colour ranges from pink to rose. The fruits are raspberry-like, with 20 to 30 drupelets, and are reddish to dark purple, very sweet, fragrant, and juicy.

Cloudberry is a low herbaceous perennial about 10 to 20 cm tall, with erect, unbranched stems, each with only 1 to 3 round-lobed leaves. The flowers are white, solitary, and terminal, with the male and female flowers on separate plants. The compound fruits have large drupelets that are yellowish, soft, and very juicy when ripe. They are somewhat sour, but delicious to those who have acquired a taste for them.

Hairy and trailing raspberries have compound leaves, the former being 3-foliate and the latter usually 5-foliate. They have a more trailing habit than arctic raspberry and cloudberry, and trailing raspberry has only 3 to 5 drupelets per fruit. None of these species have spines or thorns.

# Dwarf Raspberries

(Rose Family)

---

*Rubus chamaemorus*

***Rubus arcticus* L., *R. chamaemorus* L. and other *Rubus* species** (Rosaceae)

## Where to Find

Arctic raspberry grows in wet meadows, muskegs, and along creeks throughout most of northern Canada from British Columbia to Newfoundland. *R. arcticus* ssp. *acaulis* (Michx.) Focke is the most common sub-species. Cloudberry grows in peat bogs and muskegs throughout much of northern Canada, extending southwards in acid boggy sites. Hairy raspberry occurs throughout most of Canada in woods and thickets. Trailing raspberry is found only in the West, in western Alberta and British Columbia, where it grows in damp, montane woods.

## How to Use

Arctic raspberries are small, but so fragrant and delicious that many people consider them the choicest of all the wild fruits, surpassing even wild strawberries in flavour and bouquet. The best way to enjoy them is to pick a handful, sit down, and eat them one or two at a time, savouring every mouthful. If you do collect a sufficient number to serve as a dessert, try something simple, such as raspberry shortcake, using baking powder biscuits and cream, or raspberry whip. Do not overpower their delicate flavour by mixing them with other fruits. They freeze well, make an excellent jam, and can of course replace cultivated or wild raspberries in any recipe.

Ripe cloudberries are also excellent, especially when they mature in warm, sunny weather. Because of their tartness they are good mixed with apples or other berries. The fruits of hairy raspberry and trailing raspberry are also edible, though not as tasty as arctic raspberries and cloudberries. They can be used fresh or made into jam if enough can be collected.

## Suggested Recipes

### Arctic Raspberry Whip

| | | |
|---|---|---|
| 375 mL | arctic raspberries | 1½ cups |
| 250 mL | powdered sugar | 1 cup |
| | 1 egg white, stiffly beaten | |
| 30 mL | red wine | 2 tbsp |
| | whipped cream topping | |
| | chopped nuts (optional) | |

Combine all ingredients except whipped cream and nuts and beat well with a wire beater. Put into dessert dishes and set in refrigerator. Before serving, top with whipped cream and chopped nuts if desired. Makes 3 servings.

### Arctic Raspberry Sherbet

| | | |
|---|---|---|
| 1 L | arctic raspberries | 4 cups |
| 250 mL | sugar | 1 cup |
| 500 mL | water | 2 cups |
| 15 mL | lemon juice | 1 tbsp |
| | 1 egg white, stiffly beaten | |

Extract the juice from the berries by heating them slightly, then mashing and straining through a sieve. (The pulp can be saved and used in fruit salads or other desserts.) Boil the sugar and water for a few minutes to make a syrup. Cool, then add the raspberry and lemon juice, pour into a tray, and place in the freezer section of your refrigerator until the mixture reaches a mushy consistency. Add beaten egg white, stir, and replace in freezer. Stir mixture occasionally, and allow to freeze until firm. Serves 4.

## Eskimo Cloudberry Preserves

| | | |
|---|---|---|
| 1 L | ripe cloudberries | 4 cups |
| 250 mL | seal oil, olive oil *or* vegetable oil | 1 cup |

The traditional method of preparing these preserves is as follows: collect the ripe cloudberries in a seal-skin bag, mix them in a large bowl with seal oil, and pour the mixture into a pouch made from a seal's stomach. Place the pouch in a 3- or 4-metre-deep hole in the permafrost and cover with layers of seal skins. Thaw slightly before using. The fresh berries can also be pounded with aged caribou meat, frozen, then eaten raw, or mixed with seal oil and chewed caribou tallow and beaten to the consistency of whipped cream to make "Eskimo ice cream".

On the other hand you can collect the cloudberries in a pan, mix with olive *or* vegetable oil, put into a freezer bag, tie, and store in the freezer.

## More for Your Interest

Among those who considered the arctic raspberry the finest of all fruits was the Swedish botanist Linnaeus. He described it as superior in both smell and taste: "Its odour is of the most grateful kind, and as to its flavour, it has such a delicate mixture of the sweet and acid, as is not equalled by the best of our cultivated Strawberries" (Fernald and Kinsey 1958, p. 237). The berries of arctic raspberry and cloudberry are greatly enjoyed by Scandinavians and Laplanders and by the North American Inuit and Indians. Both are known to have a healthy vitamin C content and to also contain some vitamin A. The Scandinavians export fine cloudberry preserves, and make a delicious cloudberry liqueur that is sold around the world.

## Other Names
Red raspberry, American raspberry; well-known raspberry relatives include blackcap, or black raspberry, thimbleberry, and salmonberry.

## How to Recognize
The wild raspberry closely resembles our garden varieties and is thus easy to recognize. The woody, upright stems, usually 1 to 2 m tall, are covered with sharp prickles. The leaves are compound, with 2 to 4 lateral leaflets and 1 usually large terminal leaflet. The leaflets are pointed, and the margins sharply toothed and sometimes notched. The white-petalled flowers are borne in small clusters on short branches off the main stems. The fruits are red (or occasionally yellowish) and juicy, each an aggregate of several to many tiny individual fruits, or drupelets. When ripe the berries, unlike blackberries, separate easily from the white central receptacle and fall off in a typical "thimble" form. Wild raspberries are smaller and rounder than their cultivated counterparts, but most people agree that they are superior in flavour. The American wild raspberry is considered by many botanists to be the same species as the European wild raspberry (*R. idaeus* L.), the two being distinguished only at the variety level. Cultivated forms were developed from both European and North American types, but the American ones are hardier and more frost-resistant.

Aside from the various dwarf raspberries, treated separately in this book, there are several raspberry relatives that deserve special mention. They include: blackcap, or black raspberry (*R. leucodermis* Dougl.), a prickly shrub with arching branches, raspberry-like leaves, and dark purplish-black berries; thimbleberry (*R. parviflorus* Nutt.), an unarmed erect shrub with very large maple-leaf-like 5-lobed leaves and bright-red fruits that resemble raspberries but are softer and more shallowly cupped; and salmonberry (*R. spectabilis* Pursh), a tall, prickly shrub with 3-parted raspberry-like leaves and large, watery fruits that come in two colour forms—golden and ruby. Another species, (*R. occidentalis* L.), known

# Common Wild Raspberry and Relatives

(Rose Family)

**Rubus strigosus** Michx.
and other **Rubus** species
(Rosaceae)

as black raspberry, or *framboisier noir* in Quebec, closely resembles the western blackcap. The branches are bluish grey, long-arching, rooting at the tips, and armed with hooked prickles. The leaves are raspberry-like, the flowers white, and the fruit purply black. There are other related species in various parts of Canada. All are recognizable as raspberry relatives, and all have edible fruits that can be used in the same way as those of the true raspberry.

## Where to Find

Wild raspberry grows in moist to dry woods and on rocky slopes across Canada from Newfoundland to British Columbia, being absent only in the Pacific coastal region and above the tree line in the Far North. Blackcap grows in woods and clearings in the southern half of British Columbia. Thimbleberry grows, often in dense thickets, in open woods from British Columbia into Alberta and in the Great Lakes area of Ontario. Salmonberry is restricted to coastal British Columbia, in damp woods and moist ground, where it is locally very abundant.

## How to Use

A popular fruit throughout the Northern Hemisphere, wild raspberries can be used for making all kinds of preserves, beverages, and desserts, both raw and cooked. To our knowledge and experience wild raspberries far surpass the cultivated ones in flavour and will yield an even better jam, jelly, or preserve. They are delicious raw on any breakfast cereal, and make an excellent sauce for ice cream when heated for a very short time (overcooking may destroy some of their flavour).

Raspberries are well adapted to freezing. If you wish to sweeten them, try 1 kg (2 lb) sugar per 4 kg (8 lb) berries. They may also be dried, separately like raisins, or mashed and poured out on a sheet of heavy waxed paper to make "leather" (see p. 145). Blackcaps, thimbleberries, and salmonberries can all be substituted for wild raspberries in any recipes, although you may find it necessary to reduce the quantities of liquids used, especially for salmonberries, which are often very watery.

**Suggested Recipes**

## Wild Raspberry Bread

| 125 mL | butter | ½ cup |
|---|---|---|
| 175 mL | honey | ¾ cup |
| 125 mL | milk | ½ cup |
| | 1 egg, well beaten | |
| 500 mL | pastry flour | 2 cups |
| 2 mL | salt | ½ tsp |
| 10 mL | baking powder | 2 tsp |
| 2 mL | cream of tartar | ½ tsp |
| 2 mL | cinnamon | ½ tsp |
| | juice of 2 lemons | |
| 500 mL | fresh wild raspberries | 2 cups |

In a bowl, cream butter and honey together. Stir in milk and beaten egg. Sift in flour, salt, baking powder, cream of tartar, and cinnamon. Add lemon juice and beat until smooth. Add raspberries and pour the mixture into greased loaf pan. Bake at 180°C (350°F) for 45 minutes, or until firm and browned.

## Raspberry Summer Soup

| 500 mL | fresh wild raspberries | 2 cups |
|---|---|---|
| 250 mL | fresh cream | 1 cup |
| | few drops vanilla | |
| 25 mL | sugar | 1½ tbsp |
| 250 mL | milk | 1 cup |

Place all the ingredients into blender and blend until mixture is velvety smooth. Serve ice-cold, garnished with a few raspberries. Serves 4.

## Red Raspberry Shrub

| | | |
|---|---|---|
| 1 L | wild red raspberries | 4 cups |
| 250 mL | apple vinegar | 1 cup |
| | sugar | |

Pour vinegar over berries and let stand
4 days. Drain off juice, measure, and add
175 mL ($^3/_4$ cup) sugar to each 250 mL
(1 cup) juice. Boil gently for 15 minutes.
Cool. To serve, dilute 1 part of syrup with
3 parts of water and pour into tall tumblers
filled with crushed ice.

This traditional "hospitality" beverage,
to be at its best, should be made with old-
fashioned unpasteurized apple vinegar
(which you can make yourself or find in
specialty food stores). There is no taste of
vinegar in the syrup, but the syrup derives a
tempting sparkle and an elusive flavour from
the apples. (From Constance Conrader,
"Wild Harvest".)

## Raspberry Cooler

| | | |
|---|---|---|
| 500 mL | raspberries | 2 cups |
| 500 mL | fresh milk | 2 cups |
| | 2 to 3 drops almond extract | |
| 50 mL | sugar | $^1/_4$ cup |
| | dash of salt | |
| 250 mL | vanilla ice cream | 1 cup |

Blend together raspberries, milk, almond
extract, sugar, and salt in blender until
smooth. Add ice cream and run on low speed
until blended. Serve at once. Serves 5–7.

## Simple Raspberry Jam

| | | |
|---|---|---|
| 500 mL | fresh wild raspberries | 2 cups |
| 375 mL | sugar | 1$^1/_2$ cups |
| | juice of 1 lemon | |

Combine berries and sugar in a saucepan
and place over heat. Bring to a boil. Boil
rapidly for about 20 minutes, add lemon
juice, and boil again until jelling stage is
reached (5 to 10 minutes). Pour into hot,
sterilized jars and seal with melted paraffin.
Store in a cool place. Makes about 3
medium-sized jelly glasses.

## More for Your Interest

The young, tender sprouts of the wild raspberry can be peeled and eaten raw in the spring. The dried leaves make a very pleasant and soothing tea, made even better by the addition of a little of the mashed berries or juice. The roots can be simmered in water to produce a common folk remedy for "summer complaint"—diarrhoea and digestive upsets.

In Europe, red raspberry syrup is one of the best-known ingredients in summer drinks. Carbonated soft drinks were introduced in most European countries from America only after the Second World War. Previously, the only cooling drink was a mixture of soda water and fruit syrups, such as raspberry, orange, or lemon, served icy cold. When wild raspberries were in season many open markets in Europe featured them, and the demand for them was great. Nearly every home was equipped with a small, portable press, a device that made the extraction of juice from the berries very simple. The raw juice obtained from this press was then processed by boiling with sugar for 20 to 30 minutes. To make this syrup use 1 kg (2 lb) of sugar per 1 L (4 cups) of juice, then bottle for future use. When you want an old-style European raspberry summer drink, simply mix about 5 mL (1 tsp) of the syrup in a glass with ice-cold soda water.

Wild raspberries and their relatives were eaten, fresh and dried, by Indians throughout North America. Thimbleberries and salmonberries were, and still are, very popular among the Indian peoples of British Columbia, often being eaten with large quantities of oil. The tender young shoots of these shrubs were also eaten, and the leaves were sometimes made into tea.

**Other Names**
Dewberry, bramble.

**How to Recognize**
The many types of blackberries in Canada
are closely related to the raspberries, and,
like them, have compound (aggregate) fruits
—many small juicy fruits called drupelets,
each with a single seed, adhering together to
make up the "berry". When ripe, blackberries break off with the whitish receptacle in
the middle still attached, in contrast to
raspberries, which break off freely from the
receptacle, leaving a hollow cavity in the
middle of the fruit.

   The blackberries as a group are extremely
complex and variable. They hybridize
freely, and even botanists hesitate to classify
them into species in the ordinary sense of
the word. The fruits of all of them are edible,
but some are juicier and more flavourful
than others. Some people find it convenient
to distinguish the trailing, slender-caned
blackberries as dewberries and the stouter,
more upright ones as regular blackberries,
but this distinction is sometimes difficult to
make. The stems of all blackberries are
woody and are usually armed with slender or
stout prickles. They are usually biennial (in
some forms perennial), persisting through
the winter and bearing flowers on the second

# Wild Blackberries

(Rose Family)

*Rubus procerus*

**Rubus species**
(Rosaceae)

year (or successive years). The leaves are compound, with usually 3 to 5 pointed leaflets, coarsely to finely toothed at the edges and often with spines along the veins underneath. The flowers are white to pinkish, in small to large clusters, and the ripe berries are dark red to black, juicy, and often somewhat acid.

The most widespread native blackberry in eastern Canada is the common blackberry (*R. allegheniensis* Porter). In the West, the small but flavourful trailing wild blackberry, or western dewberry, (*R. ursinus* Cham. & Schlecht), is very common in parts of southern British Columbia. In many areas, particularly on Vancouver Island and the lower mainland of British Columbia, but also in eastern Canada, two very aggressive Old World species, Himalaya blackberry (*R. procerus* P.J. Muell.) and cut-leaved blackberry (*R. laciniatus* Willd.), have escaped from cultivation and are very prevalent.

## Where to Find

Blackberries and dewberries are mainly restricted to the eastern provinces and southern British Columbia. The native blackberries grow in dry, open woods and along roadsides. The introduced ones grow along roadsides and in waste places, old orchards and gardens, where they form dense, impenetrable thickets and often choke out other less aggressive plants.

## How to Use

Wild blackberries and dewberries, among the finest of all our wild fruits, can be eaten fresh or cooked, alone or with sugar, cream, or other dressings. Commercially, blackberries are used to flavour ice cream and to make syrups and beverage concentrates. They make delicious jams, jellies, and preserves, and are good in combination with apples, raspberries, and other fruits. They also make an excellent wine.

Pies and tarts made from blackberries are thought by many to be without equal. One of the best pies Adam Szczawinski ever tasted was made by Scottish friends during a camping trip in blackberry country. Christened "bramble delight", this pie was easy to make. His friends simply collected ripe blackberries, dumped them into a deep pie dish, sprinkled them with sugar, cornstarch, cinnamon, and a few dabs of margarine, and covered them with a thick layer of pastry made with flour, soft margarine, and a little water and salt. The pie was baked in a moderate oven (in this case a makeshift campfire oven) until the crust was golden brown (about 45 minutes). This delicious pie was made without the aid of a single measuring device. If you are fortunate enough to have blackberries growing near your camping place or your home, do not let them go to waste. Try some of our recipes or experiment with your own.

## Suggested Recipes

### Blackberry Sauce (for Cheesecake)

| 500 mL | blackberries | 2 cups |
|---|---|---|
| 50 mL | cornstarch | ¼ cup |
| 50 mL | sugar | ¼ cup |
| 30 mL | butter | 2 tbsp |
| 15 mL | lemon juice | 1 tbsp |
| 50 mL | port wine | ¼ cup |

Press berries through a sieve to remove seeds. Mix cornstarch and sugar in saucepan. Stir in the sieved blackberries and cook, stirring constantly until mixture boils. Remove from heat, add butter, and stir until melted. Add lemon juice and port wine. Chill. Makes about 500 mL (2 cups) cheesecake sauce.

### Bramble Jelly

| 1 L | blackberries | 4 cups |
|---|---|---|
| 250 mL | water | 1 cup |
| 5 mL | tartaric *or* citric acid | 1 tsp |
| | sugar | |

Wash berries and place in pan with water and acid and simmer for 1 hour. Strain through a jelly bag overnight, or until dripping stops. Measure juice, and add sugar in the proportion of 250 mL (1 cup) sugar to 250 mL (1 cup) juice. Return juice and sugar to pan and boil until jelly sets when tested on a cold plate. Pour into hot, sterilized jars and seal with melted paraffin. Store in a cool place. Yields 4 medium-sized glasses.

## Blackberry Wine

| | | | |
|---|---|---|---|
| 5 L | blackberries | 4 | qt |
| 5 L | boiling water | 4 | qt |
| | sugar | | |
| 8 g | dry yeast (1 package) | 1/4 | oz |
| | 1 slice toast | | |

In a crock or other large container cover berries with water, allowing them to stand for 24 hours, then bring to a boil, let boil for 5 minutes, and allow to cool. Strain through a jelly bag or very fine sieve. Add sugar in proportion of 1 volume sugar to 2 of juice. Return to crock, float yeast on a slice of dry toast on top, and let stand for 3 weeks at room temperature (preferably in darkness). Strain again and bottle. Cork lightly for a few days, and when fermentation (bubbling) ceases, tighten corks and store in a dark place for 6 to 12 months before drinking. Blackberries make one of the best wines, good in flavour and body. The berries can also be mixed in a 2-to-1 proportion with Italian plums to make a pleasant wine.

## More for Your Interest

North American wild blackberries have been used in developing a number of different cultivated varieties now very popular commercially. These include loganberry, boysenberry, youngberry, and cascadeberry, as well as garden blackberry strains.

Native peoples throughout the continent enjoyed eating wild blackberries and dewberries. They used them fresh and also mashed them into cakes and dried them in the sun or over a fire for winter use. Sometimes the berries were dried whole, like raisins. The leaves, especially the old, brownish ones, were occasionally used to make a tea.

## Other Name
Rowan tree.

## How to Recognize
There are several species of native mountain-ash in Canada, including two eastern species that are small trees—American mountain-ash (*S. americana* Marsh.) and showy mountain-ash [*S. decora* (Sarg.) Schneid.]—and two western species that are more shrub-like—Sitka mountain-ash (*S. sitchensis* Roemer) and western mountain-ash (*S. scopulina* Greene). In addition, the European mountain-ash, or rowan tree, which grows up to 15 m in height, is widely planted as an ornamental in Canada; in some places it has escaped cultivation and is found growing wild.

All of these species are deciduous, with alternate, pinnately compound leaves up to 25 cm long, but usually shorter. These are composed of 7 to 17 similar leaflets in opposite pairs along the main leafstalk, with one at the end. The leaflets are narrowly oval-shaped, usually sharp-tipped, with teeth along the margins (at least around the upper end). The bark is thin, smooth, and light grey, sometimes becoming scaly on the tree species when mature.

The flowers are small and white in many-flowered, flat-topped, or rounded clusters. Like apples, the fruits are classed as pomes, but are berry-like in size and appearance. They are soft, mealy, bright orange to red, and bitter-tasting, often remaining on the branches after the leaves fall. The fruit is considered low in protein but high in carbohydrates, and has a considerable tannin content.

## Where to Find
The mountain-ashes grow in many different soils and under a variety of conditions. American mountain-ash is found from central Ontario and Quebec to Newfoundland and the Maritimes southwards. The range of showy mountain-ash is similar, but extends westwards to central Manitoba and farther north in Ontario and Quebec. The two western species are largely montane, extending from western Alberta to the Pacific coast. The European mountain-ash can be found throughout southern Canada, usually close to habitations.

# Mountain-Ashes

(Rose Family)

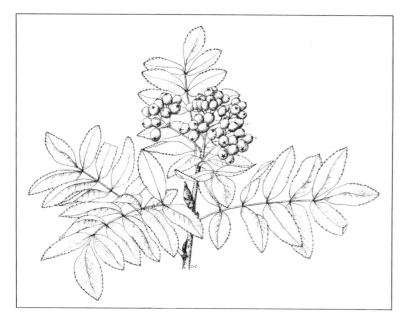

## How to Use

When eaten raw, the berries are rather bitter due to the high tannin content. If you would like to try them raw, or even if you plan to cook them, wait until they have been exposed to frost, when they taste much better. In Europe, rowan jelly is well known and is preferred by many to red currant jelly. It is slightly bitter in taste and usually eaten with roast mutton, venison, and hare. The most common fruit combinations for the berries are crabapples and citrus fruits. Together these make good jellies, marmalades, and jams. A variety of liquors can be produced from mountain-ash berries. Some of them, such as rowan brandy, made in Poland, are world famous. Mountain-ash wines and liqueurs are also available on European markets and are becoming more popular in North America.

## Warning

Mountain-ash berries should always be fully ripe when used. The seeds contain amygdalin, like those of the *Prunus* species, and therefore mountain-ash jam should be used sparingly. As the fruits are also high in tannin, they should be used in moderation (see Warnings for sumac and oak, pp. 32 and 102).

*Sorbus sitchensis*

## Suggested Recipes

### Rowan Jelly

| 2 kg | mountain-ash fruits | 4 lb |
|---|---|---|
| 500 mg | apples *or* crabapples | 1 lb |
| 5 mL | citric *or* tartaric acid | 1 tsp |
| | water | |
| | sugar | |

Wash the fruits, cutting the apples in pieces without peeling or coring. Place with the citric *or* tartaric acid in a large saucepan and add just enough water to cover the fruit. Simmer 45 minutes to 1 hour, until the fruits are pulpy, adding more water if necessary. Strain for several hours through a fine sieve or jelly bag. Measure juice and return to kettle with sugar in the proportion of 500 mg (1 lb) sugar to each 500 mL (2 cups) juice. Stir until sugar is dissolved. Boil briskly 10 to 15 minutes until the jelly sets when tested on a cold plate, then pour into sterilized jelly glasses and seal with melted paraffin when cool. Store in a cool, dark place. Makes about 15 medium-sized jelly glasses.

### Mountain-Ash Jam with Ginger

| 1.5 kg | mountain-ash fruits | 3 lb |
|---|---|---|
| 1.5 kg | sugar | 3 lb |
| 100 g | preserved ginger | 4 oz |
| | 2 large lemons | |
| | 1 large orange | |

Wash mountain-ash fruits and place in a large saucepan. Add enough water to barely cover the fruit, and boil until tender. Rub through a sieve. Add the ginger, cut into very small pieces, and the juice and grated rind of the lemons and orange. Return to pan and bring to a boil. Boil briskly for 20 minutes, or until the jam sets when dropped onto a cold plate. Pour into sterilized jars and seal with melted paraffin when cool. Store in a cool, dark place. Makes about 8 medium-sized glasses.

## Mountain-Ash Brandy-Wine

| | | |
|---|---|---|
| 4.5 L | mountain-ash fruits | 1 gal |
| 4.5 L | lukewarm water | 1 gal |
| 5 kg | coarse brown (demerara) sugar | 8½ lb |
| | juice of 5 lemons | |
| | juice of 5 oranges | |
| 8 g | dry yeast (1 package) | ¼ oz |
| | 1 slice toast | |

Mix all ingredients except yeast and toast in a 25-L (5-gal) crock. Put yeast on a piece of toast and float on the top to dissolve slowly. Do not put a lid on, although you may cover with a cloth. Put crock in a warm place where the temperature is constant, and after 3 days remove the toast. Leave the crock to work, stirring daily for 10 to 12 days, when it should be finished foaming, then strain. Wash the crock and put the strained mixture back in it. Leave another 2 days, then strain again and bottle, but do not cork tightly until fermentation stops (about 3 more days). Strain off again, rebottle, and cork firmly. Age for at least 6 months, preferably 1 year, before using.

## More for Your Interest

In the Scottish Highlands and in Wales, the juice of the European rowan is fermented into a liquor resembling cider. In various parts of northern Europe in times of scarcity, the dried fruits were ground into a meal and used to make bread. In British Columbia, the Thompson and Haida Indians ate mountain-ash fruits, often with fish or meat.

When canning blueberries, try putting a cluster of mountain-ash fruits in the top of each jar as a flavouring.

## Other Names

*V. riparia* is known as frost or riverbank grape, and *V. aestivalis* is called summer, pigeon, or bunch grape.

## How to Recognize

The wild grapes are trailing or climbing deciduous vines, closely resembling the various cultivated types. There are over a dozen species in eastern North America, but only two occur commonly in Canada. Another species, the fox grape (*V. labrusca* L.), from which many of our cultivated strains are derived, is occasionally found as an escape. All of the wild grapes are similar. The vines cling to other vegetation by means of strong tendrils borne along the stems. The leaves are broadly heart-shaped or round in outline, with 3 to 5 pointed lobes and sharply toothed margins. The young leaves of the frost grape are covered with short, straight, whitish hairs. The young leaves of the summer grape have long, cobwebby, rust-coloured hairs underneath. The flowers of the grapes are greenish and inconspicuous, in compact, pyramidal clusters. These ripen into dark, spherical berries that are tart but tangy. Those of the frost grape are usually about 10 to 12 mm across, those of the summer grape slightly smaller. Fox grape berries range from 1 to 2 cm in diameter.

## Where to Find

The wild grapes grow in moist woods, along roadsides, and in thickets. The frost grape occurs commonly in Canada from New Brunswick and Quebec to Manitoba. The summer grape is restricted in Canada to southern Ontario.

## How to Use

The fruit of all the wild grapes is edible, although the berries tend to be more tart than those of cultivated grapes. They are very flavourful, however, and are good for making juice, jellies, jams, conserves, and pies. They also make excellent raisins for baking and nibbling. Like commercial raisins, they are high in iron. Sometimes the grapes will turn to raisins right on the vines and can be gathered ready-made. Wild grape preserves are more fragrant and better tasting than those from cultivated grapes. Wild grapes can also be used for winemaking, but because of their lower sugar content, especially in the northern part of their range, they are not considered as good for wine as cultivated types.

# Wild Grapes

(Grape Family)

*Vitis riparia* **Michx.**
**and** *V. aestivalis* **Michx.**
(Vitaceae)

Wild grapes are improved by exposure to frost, and are best if left until fully ripe before harvesting. However, if you wish to make grape jelly without using commercial pectin, use about half ripe and half unripe grapes.

Grape leaves are popular in southern Europe and Middle Eastern countries, where they are used in many traditional dishes. The leaves have a pleasant acid flavour and blend very well with meat and rice. They are said to be best for cooking in late spring, before they become too tough.

## Warning

The Canadian moonseed (*Menispermum canadense* L.) is a plant unrelated but superficially similar in appearance to the wild grape. The fruits resemble wild grapes but are very poisonous, containing a nerve poison with curare-like action. (Curare was used by the Indians of South America to make poisoned arrows.) Moonseed fruits have a single seed, whereas grapes have several seeds and the moonseed leaves have smooth edges, whereas those of wild grapes are sharply toothed.

## Suggested Recipes

### Wild Grape Raisins

Gather as many wild grapes as you can and spread them out evenly in a single layer on a framed wire or fibreglass screen. Cover with cheesecloth to keep flies away and place the frame out in the sun. Each evening the frame should be taken inside and kept overnight in a warm place. The drying process should take about 3 days, depending on the weather and on the size of the grapes. The raisins should be moist and soft to touch, and should be stored in covered containers under refrigeration.

### Wild Grape Juice

| 4.5 kg | wild grapes | 10 lb |
|--------|-------------|-------|
|        | water to cover | |
| 1 kg | sugar | 2 lb |

De-stem and thoroughly wash grapes. Crush fruit in a preserving pan, but do not crush seeds. Add just enough water to cover the fruit and simmer for 30 minutes. Press the fruit through a sieve, then strain through a jelly bag or cheesecloth. Add sugar to taste and simmer for 15 minutes. Pour immediately into hot, sterilized jars and seal. Store in a dark, cool place. Makes about 3 L (2$\frac{1}{2}$ qt).

## Wild Grape Jelly

| | | |
|---|---|---|
| 1 L | ripe wild grapes | 4 cups |
| 1 L | unripe wild grapes | 4 cups |
| | juice and peel from 1 orange | |
| | sugar | |

De-stem and wash grapes, then place in a preserving pan with the orange juice and finely chopped peel. Mash the fruit without crushing the seeds. Simmer on low heat for about 30 minutes, or until fruit is completely soft. Let drip through jelly bag overnight. In the morning measure juice and return to saucepan. Measure out an equal volume of sugar and heat in a slow oven at 120°C (250°F) until just hot enough to touch. Meanwhile, bring juice to a boil again, and after boiling 5 minutes add the hot sugar. Bring to a boil quickly and boil hard for 5 additional minutes, stirring frequently. When the liquid jells on a cold plate or forms two drips off the edge of a wooden spoon, remove from heat and bottle immediately in hot, sterilized jars. Seal with hot paraffin and store in a cool, dark place. Yields about 8 medium-sized jelly glasses.

## More for Your Interest

Wild grape stems yield a sweet, watery sap that can be drunk when water is not available. Cut the vine off near ground level and place the severed end in a container. Then make a slantwise cut in the stem about 2 m up to enable the sap to drain out into the container. Indian people in the Midwest have used this beverage for centuries.

The European grape, *V. vinifera* L., has been under cultivation for over 4,000 years, and its use in making wine was detailed as early as the Fourth Dynasty in Egypt (2440 B.C.). Vineyards and wine are also mentioned in the biblical story of Noah. The European species was brought to the New World by Columbus, and by 1741 there were said to be thousands of the vines from Portugal thriving in Georgia. Today, it is grown extensively in California and in the Okanagan Valley of British Columbia. East of the Rocky Mountains, most of the cultivated grapes are derived from native American species or are hybrids of native species with the European grape.

Glossary
Bibliography
Index

# Glossary

**Achene**
Small, dry, indehiscent, one-seeded fruit.

**Alternate**
Any arrangement of leaves or other parts not opposite or whorled. Borne singly at successive nodes.

**Annual**
A plant that lives only one growing season.

**Anther**
The pollen-bearing portion of the stamen.

**Axis**
The central line of any organ or support of a group of organs, such as a stem.

**Basal**
Pertaining to the base or lower parts of a plant or structure.

**Berry**
The most generalized type of fleshy fruit, developed from a single pistil and containing several seeds.

**Biennial**
A plant that lives only two years; flowers and fruits are usually produced in the second year.

**Blade**
The expanded, usually flat portion of a leaf.

**Bloom**
A waxy powder covering a surface, such as the skin of a fruit, making it glaucous.

**Bract**
A modified leaf, either small and scale-like or large and petal-like.

**Calyx**
The outermost whorl of floral parts on a flower; made up of sepals.

**Cambium**
A slimy layer between the wood and the bark of trees and shrubs from which new wood and bark tissues are derived.

**Capsule**
A dry fruit, composed of more than one carpel, that splits open when ripe.

**Carpel**
A single pistil or one unit of a compound pistil.

**Caryopsis**
A one-seeded fruit of grasses, with the seed coat adhering to the outer wall of the fruit.

**Catkin**
A drooping, elongated cluster of petalless flowers, either male or female, as in willows, alders, and birches.

**Circumboreal**
Occurring all the way around the North Pole, in both Northern Eurasia and North America.

**Compound**
Composed of two or more similar parts, as a compound leaf or fruit.

**Crown**
The leafy or branching head of a tree.

**Deciduous**
A plant that sheds all its leaves annually, as opposed to being evergreen.

**Dehiscent**
Splitting open at maturity by means of slits or valves.

**Dicotyledons**
A large group of flowering plants characterized by having embryos with two seed leaves (cotyledons), net-veined leaves, and flower parts in fours or fives (as opposed to monocotyledons).

**Disc**
A fleshy expansion of the receptacle.

**Drupe**
A fleshy fruit with a hard inner stone, such as a cherry or plum.

**Drupelet**
One segment of an aggregate fruit, such as raspberry or blackberry.

**Elliptical**
Having the shape of an ellipse; a compressed circle.

**Evergreen**
A plant that has green leaves throughout the year, as opposed to being deciduous.

**Family**
A category in the classification of plants and animals, ranking above a genus and below an order; includes one genus, or two or more related genera.

**Fruit**
A ripened seed-case or ovary and any associated structures that ripen with it.

**Fruit leather**
Sheets of dehydrated mashed fruit, similar to leather in consistency.

**Genus** (plural: genera)
A category in the classification of plants and animals, it is the main subdivision of a family; includes one species, or two or more related species.

**Glabrous**
Smooth, not hairy.

**Gland**
A secreting surface or structure; any appendage having the appearance of such an organ.

**Glandular**
Bearing glands or secreting hairs.

**Glaucous**
Whitened; covered with white bloom.

**Habit**
The general appearance of a plant.

**Habitat**
The situation in which a plant grows.

**Head**
A dense cluster of sessile or nearly sessile flowers or fruits on a very short axis or receptacle.

**Herb**
A plant with no woody stem above ground level; dies back to ground level every year.

**Herbaceous**
A herb; not woody.

**Hybrid**
A cross between two species.

**Indehiscent**
Not splitting open at maturity (pertaining to dry fruits).

**Lanceolate**
Considerably longer than broad, tapering upwards from the middle or below; lance-shaped.

**Leaflet**
An ultimate unit of a compound leaf.

**Lenticel**
A slightly raised area in the bark of the stem or roots of a tree or shrub, through which gaseous exchange occurs between the stem tissues and the atmosphere.

**Lobed**
Having major divisions extending about halfway to the base or centre; often applied to leaves, such as oak or maple.

**Monocotyledons**
A large group of flowering plants characterized by having embryos with a single seed leaf (cotyledon), parallel-veined leaves, and flower parts in threes (as opposed to dicotyledons).

**Muskeg**
A poorly drained area with acid conditions, characterized by the presence of sphagnum moss; a common landscape feature in northern Canada.

**Node**
A joint or portion of a stem from which a leaf or branch arises.

**Nut**
A hard, (usually) one-seeded, indehiscent fruit.

**Nutlet**
A small nut.

**Oblong**
Longer than broad, with the margins nearly parallel (pertaining to leaves).

**Opposite**
Growing directly across from each other at the same node (in reference to leaves).

**Orbicular**
Circular in outline.

**Ovary**
The part of the pistil (female flower organ) that contains the ovules, or immature seeds.

**Ovoid**
Egg-shaped.

**Ovule**
An immature seed.

**Palmate**
Having lobes radiating from a common point; resembling a hand with fingers spread.

**Perennial**
A plant that lives more than two years.

**Pericarp**
The outer wall of a fruit.

**Petiole**
A leafstalk.

**Pinnate**
Compound, with leaflets arranged in feather-like fashion on either side of the axis or petiole (pertaining to leaves).

**Pistil**
The female organ of a flower.

**Pollen**
The male fertilizing grain produced by the anther.

**Pome**
An apple-like fleshy fruit.

**Rachis**
The axis of a spike or a compound leaf.

**Ray**
The petal-like marginal flowers of sunflower, daisy, or other composite flowerheads.

**Receptacle**
The more-or-less expanded summit of the stalk which bears a flower or flower head.

**Recurved**
Curving or curling backwards.

**Rhizome**
A creeping underground stem or rootstalk, serving in vegetative reproduction and food storage; distinguished from a true root by the presence of nodes, buds, or scale-like leaves.

**Runner**
A slender stolon, rooting at the nodes or tip.

**Scale**
Modified leaf, often papery or woody, as on the cones of conifers.

**Sepal**
One of the modified leaves of a calyx.

**Shrub**
A woody plant, creeping or upright, that is smaller than a tree and usually has multiple stems.

**Species**
The fundamental unit in the classification of plants and animals.

**Spike**
An elongated flower cluster, with flowers attached directly to the central stalk.

**Spine**
A sharp, woody outgrowth of the stem.

**Stamen**
The male organ of a flower.

**Stolon**
A sucker runner or any basal branch of a plant that is disposed to root.

**Stoloniferous**
Tending to produce stolons or runners.

**Tendril**
A slender coiling or twining organ
growing from a stem or leaf.

**Terminal**
Growing at the end of a stem or
branch.

**Thorn**
A stiff, woody modified stem or
branch with a sharp point.

**Tree**
A large, usually single-stemmed
woody plant.

**Unarmed**
Without thorns, spines, or prickles.

**Whorl**
Leaves or other organs arranged
very close together at a node or at
the base of a stem.

# Bibliography

**Adney, E. Tappan**
(1944). "The Malecite Indian's Names for Native Berries and Fruits, and Their Meanings". *Acadian Naturalist*, Vol. 1, No. 3, pp. 103–10.

**Anderson, Jacob Pet**
(1939). "Plants Used by the Eskimos of the Northern Bering Sea and Arctic Regions of Alaska". *American Journal of Botany*, Vol. 26, pp. 714–16.

**Anderson, James R.**
(1925). *Trees and Shrubs, Food, Medicinal and Poisonous Plants of British Columbia*. Victoria: British Columbia Department of Education.

**Angier, Bradford**
(1972). *Feasting Free on Wild Edibles*. Harrisburg, Pa.: Stackpole.
(1974). *Field Guide to Edible Wild Plants*. Harrisburg, Pa.: Stackpole.

**Assiniwi, Bernard**
(1972). *Recettes indiennes et survie en forêt*. Montreal: Leméac.

**Bean, Lowell John, and Katherine Saubel**
(1972). *Temalpakh (from the Earth): Cahuilla Indian Knowledge and Usage of Plants*. Morongo Indian Reservation, Banning, Calif.: Malki Museum Press.

**Berglund, Berndt, and Clare E. Bolsby**
(1974). *The Edible Wild*. Toronto: Modern Canadian Library.

**Black, Marmelade**
(1977). *It's the Berries*. Saanichton, B.C.: Hancock House.

**Boorman, Sylvia**
(1962). *Wild Plums in Brandy*. Toronto: McGraw-Hill.

**Brackett, Babette, and Maryann Lash**
(1975). *The Wild Gourmet*. Boston: David R. Godine.

**Brown, D.K.**
(1954). *Vitamin, Protein, and Carbohydrate Content of Some Arctic Plants from the Fort Churchill, Manitoba, Region*. Defence Research Northern Laboratory, Technical Paper 23. Ottawa: Defence Research Board.

**Budd, Archibald C., and Keith F. Best**
(1964). *Wild Plants of the Canadian Prairies*. Canada Department of Agriculture, Publication 983. Ottawa: Queen's Printer.

**Calder, James A., and Roy L. Taylor**
(1968). *Flora of the Queen Charlotte Islands*. Canada Department of Agriculture, Research Branch Monograph No. 4, Pt. 1. Ottawa: Queen's Printer.

**Canada Department of Agriculture**
(1936). *Jams, Jellies and Pickles*. Publication 535. Ottawa: King's Printer.
(1943). *Wartime Canning*. Publication 751. Ottawa: King's Printer.
(1972). *Making Blueberry Wine at Home*. Publication 1206. Ottawa.
(1975). *Home Preparation of Juices, Wines and Cider*. Publication 1406. Ottawa: Supply and Services Canada.
(1976). *Jams, Jellies and Pickles*. Publication 992. Ottawa: Supply and Services Canada.

**Canada Department of Health and Welfare**
(1971). *Indian Food: A Cookbook of Native Foods from British Columbia*. Vancouver: Medical Services Branch, Pacific Region.

**Canada Department of Indian Affairs and Northern Development**
(1972). *Northern Survival*. Ottawa: Information Canada.

**Clark, Lewis J.**
(1973). *Wild Flowers of British Columbia*. Sidney, B.C.: Gray's.

**Claus, Edward P., Varro E. Tyler, and Lynn R. Brady**
(1970). *Pharmacognosy*. Philadelphia: Lea and Febiger.

**Conrader, Constance**
(1964–65). "Wild Harvest".
*Wisconsin Trails.* Pt. 1: Vol. 5, No.
2, pp. 27–32; Pt. 2: Vol. 6, No. 2,
pp. 15–19.

**Cutright, Paul Russell**
(1969). *Lewis and Clark:
Pioneering Naturalists.* Urbana:
University of Illinois Press.

**Densmore, Frances**
(1927). "Uses of Plants by the
Chippewa Indians". Pages
275–397 in *Bureau of American
Ethnology, 44th Annual Report,
1926–27.* Washington, D.C.:
Smithsonian Institution.

**Drury, H.F., and S.G. Smith**
(1956). "Alaskan Wild Plants as an
Emergency Food Source". Pages
155–59 in *Science in Alaska.*
Proceedings of the Fourth Alaskan
Science Conference, 1953. Juneau,
Alaska.

**Ellerhoff, Kay**
(1975). "Cooking the Wild Berry".
*Montana Outdoors,* Vol. 6, No. 4,
pp. 50–53.

**Ellis, Eleanor A.**
(1968). *Northern Cookbook.* Canada
Department of Indian Affairs and
Northern Development. Ottawa:
Queen's Printer.

**Fernald, Merritt L., and Alfred
C. Kinsey**
(1958). *Edible Wild Plants of
Eastern North America.* Rev. by
Reed C. Rollins. New York: Harper
and Row.

**Frankton, Clarence, and Gerald
A. Mulligan**
(1970). *Weeds of Canada.* Rev. ed.
Canada Department of Agriculture,
Publication 948. Ottawa: Queen's
Printer.

**Fried, Barbara R.**
(1962). *The Berry Cookbook.* New
York: Collier Books.

**Furlong, Marjorie, and Virginia
Pill**
(1974). *Wild Edible Fruits and
Berries.* Healdsburg, Calif.:
Naturegraph.

**Gaertner, Erika E.**
(1962). "Freezing, Preservation
and Preparation of Some Edible
Wild Plants of Ontario". *Economic
Botany,* Vol. 16, No. 4,
pp. 264–65.
(1967). *Harvest Without Planting.*
Chalk River, Ont.: The author.

**Garman, E.H.**
(1963). *Pocket Guide to Trees and
Shrubs in British Columbia.* British
Columbia Provincial Museum
Handbook No. 31. Victoria.

**Garrett, Blanche Pownall**
(1975). *A Taste of the Wild.*
Toronto: Lorimer.

**Gates, Charles M., ed.**
(1965). *Five Fur Traders of the
Northwest.* St. Paul: Minnesota Historical Society.

**Gerard, John**
(1975). *The Herbal or General
History of Plants.* Repr. of the 1633
ed., as rev. and enl. by Thomas
Johnson. New York: Dover.

**Gibbons, Euell**
(1962). *Stalking the Wild
Asparagus.* New York: McKay.
(1966). *Stalking the Healthful
Herbs.* New York: McKay.

**Gillespie, William H.**
(1959). *A Compilation of the Edible
Wild Plants of West Virginia.* New
York: Scholar's Library.

**Gleason, H.A.**
(1952). *The New Britton and Brown
Illustrated Flora of Northeastern
United States and Adjacent Canada.*
New York: New York Botanical
Garden.

**Gordon, Eva. L.**
(1943). *Wild Foods.* Ithaca, N.Y.:
Cornell Rural School Leaflet,
Vol. 36, No. 4.

**Gray, Asa**
(1970). *Manual of Botany.* New York: Van Nostrand. Corrected printing of 8th (centennial) ed., 1950, as rev. and enl. by M.L. Fernald. New York: American Book Co.

**Hall, Alan**
(1976). *The Wild Food Trailguide.* New York: Holt, Rinehart and Winston.

**Harlow, William M.**
(1959). *Fruit Key and Twig Key to Trees and Shrubs.* New York: Dover. Repr. of 1st ed. of *Fruit Key to Northeastern Trees* and 4th rev. ed. of *Twig Key to the Deciduous Woody Plants of Eastern North America.*

**Harrington, Harold David**
(1967). *Edible Native Plants of the Rocky Mountains.* Albuquerque: University of New Mexico Press.

**Hart, Jeff**
(1976). *Montana—Native Plants and Early Peoples.* Helena, Mont.: Montana Historical Society and Montana Bicentennial Administration.

**Hearne, Samuel**
(1911). *A Journey from Prince of Wales's Fort in Hudson's Bay to the Northern Ocean in the Years 1769, 1770, 1771 and 1772.* New ed. with introd., notes, and illus. by J.B. Tyrrell. Toronto: Publications of the Champlain Society, Vol. 6.

**Heiser, Charles B., Jr.**
(1951). "The Sunflower Among the North American Indians". *Proceedings of the American Philosophical Society,* Vol. 95, No. 4, pp. 432–48.

**Hellson, John C., and Morgan Gadd**
(1974). *Ethnobotany of the Blackfoot Indians.* Ottawa: National Museums of Canada, National Museum of Man Mercury Series, Canadian Ethnology Service Paper No. 19.

**Hitchcock, C. Leo, Arthur Cronquist, Marion Ownbey, and J.W. Thompson**
(1955–69). *Vascular Plants of the Pacific Northwest.* 5 parts. Seattle: University of Washington Press.

**Hopkins, Milton**
(1942). "Wild Plants Used in Cookery". *Journal of the New York Botanical Garden,* Vol. 43, No. 507, pp. 71–76.

**Hosie, R.C.**
(1969). *Native Trees of Canada.* 7th ed. Ottawa: Queen's Printer.

**Hultén, Eric**
(1968). *Flora of Alaska and Neighboring Territories.* Stanford, Calif.: Stanford University Press.

**Jesperson, Ivan F.**
(1974). *Fat-Back and Molasses: A Collection of Favourite Old Recipes from Newfoundland and Labrador.* St. John's, Nfld.: The author.

**Kinderlehrer, Jane**
(1971). *Confessions of a Sneaky Organic Cook.* New York: New American Library.

**Kingsbury, John M.**
(1964). *Poisonous Plants of the United States and Canada.* Englewood Cliffs, N.J.: Prentice-Hall.
(1965). *Deadly Harvest: A Guide to Common Poisonous Plants.* New York: Holt, Rinehart and Winston.

**Kirk, Donald R.**
(1975). *Wild Edible Plants of the Western United States.* 2nd ed. Healdsburg, Calif.: Naturegraph.

**Knap, Alyson Hart**
(1975). *Wild Harvest: An Outdoorsman's Guide to Edible Wild Plants in North America.* Toronto: Pagurian Press.

**Leechman, Douglas**
(1943). *Vegetable Dyes from North American Plants.* Toronto: Oxford University Press.
(1949). "Don't Eat Those Berries". *Forest and Outdoors,* Vol. 45, No. 6, pp. 13, 29–31. Pts. 2–4 publ. under the title "Edible Wild Berries". Pt. 2: Vol. 45, No. 7, pp. 14–15; Pt. 3: Vol. 45, No. 8, pp. 20–21; Pt. 4: Vol. 45, No. 9, pp. 16–17.

**Lewis, Walter H., and Memory P.F. Elvin-Lewis**
(1977). *Medical Botany: Plants Affecting Man's Health.* New York: John Wiley & Sons.

**Link, Mike**
(1976). *Grazing: The Minnesota Wild Eater's Food Book.* Bloomington, Minn.: Voyageur Press.

**Mabey, Richard**
(1975). *Food for Free.* London: Fontana-Collins.

**Mallory, Megan**
(1972). *Cooking in the Orchard.* North Hollywood, Calif.: Gala Books.

**Marie-Victorin, Frère**
(1964). *Flore laurentienne.* 2nd ed., rev. by Ernest Rouleau. Montreal: Presses de l'Université de Montréal.

**Martin, Alexander C., Herbert S. Zim, and Arnold L. Nelson**
(1961). *American Wildlife and Plants.* New York: Dover.

**Masefield, G.B., M. Wallis, S.G. Harrison, and B.E. Nicholson**
(1969). *Oxford Book of Food Plants.* Oxford: Oxford University Press.

**McIlwraith, T.F.**
(1948). *The Bella Coola Indians.* 2 vols. Toronto: University of Toronto Press.

**Medsger, Oliver P.**
(1972). *Edible Wild Plants.* New York: Collier-Macmillan.

**Millspaugh, Charles F.**
(1974). *American Medicinal Plants.* New York: Dover. Repr. of 1892 ed., entitled *Medicinal Plants*, 2 vols. Philadelphia: J.C. Yorston.

**Mohney, Russ**
(1975). *Why Wild Edibles?* Seattle: Pacific Search.

**Mooders, Robert L.**
(1968). *Getting Started in Home Winemaking.* Webster Groves, Mo.: The author.

**Morton, Julia F.**
(1963). "Principal Wild Food Plants of the United States excluding Alaska and Hawaii". *Economic Botany*, Vol. 17, No. 4, pp. 319–30.

**Morwood, William**
(1973). *Traveller in a Vanished Landscape.* London: Gentry Books.

**Nichols, Nell B., ed.**
(1964). *Freezing and Canning Cookbook.* New York: Belmont-Tower.

**Nickey, Louise K.**
(1976). *Cookery of the Prairie Homesteader.* Beaverton, Oreg.: Touchstone Press.

**Oswalt, Wendell H.**
(1957). "A Western Eskimo Ethnobotany". *Anthropological Papers of the University of Alaska*, Vol. 6, No. 1, pp. 17–36.

**Palmer, Edward**
(1878). "Plants Used by the Indians of the United States". *American Naturalist*, Vol. 12, No. 9, pp. 593–606; No. 10, pp. 646–55.

**Parke, Gertrude**
(1965). *Going Wild in the Kitchen.* New York: McKay.

**Peterson, Maude Gridley**
(1973). *How to Know Wild Fruits.* New York: Dover. Repr. of 1905 ed. New York: Macmillan.

**Pokorný, J.**
(1974). *Flowering Shrubs.* London: Octopus.

**Porsild, A.E.**
(1937). *Edible Roots and Berries of Northern Canada.* Canada Department of Mines and Resources, National Museum of Canada. Ottawa: King's Printer.
(1945). *Emergency Food in Arctic Canada.* National Museum of Canada, Special Contribution No. 45-1 (mimeographed). Ottawa.
(1953). "Edible Plants of the Arctic". *Arctic*, Vol. 6, No. 1, pp. 15–34.

**Preston, Richard J.**
(1961). *North American Trees.* Rev. ed. Cambridge, Mass.: MIT Press.

**Rodahl, Kaare**
(1950). "Arctic Nutrition".
*Canadian Geographical Journal*,
Vol. 40, No. 2, pp. 52–60.

**Rombauer, Irma S., and
Marion R. Becker**
(1975). *The Joy of Cooking*.
Indianapolis, Ind.: Bobbs-Merrill.

**Rousseau, Jacques**
(1946–48). "Notes sur l'ethnobotanique d'Anticosti". *Memoirs of
the Montreal Botanical Garden*, No.
2, pp. 5–16. Repr. from *Archives
de Folklore*, No. 1 (1946), pp.
60–71.
(1946–48). "Ethnobotanique abé-
nakise". *Memoirs of the Montreal
Botanical Garden*, No. 2, pp.
17–54. Repr. with minor rev. from
*Archives de Folklore*, No. 2 (1947),
pp. 145–82.

**Rousseau, Jacques, and Marcel
Raymond**
(1945). *Études ethnobotaniques
québécoises*. Contributions de
l'Institut botanique de l'Université
de Montréal, No. 55.

**Saskatoon, City of**
(1977). *Municipal Manual*, Section
1, Saskatoon, Sask.

**Saunders, Charles F.**
(1976). *Edible and Useful Wild
Plants of the United States and
Canada.* New York: Dover. Repr.
of 1934 ed., entitled *Useful Wild
Plants of the United States and
Canada*. New York: Robert M.
McBride.

**Scoggan, H.J.**
(1978–79). *Flora of Canada*. Pts.
1–3 (1978); Pt. 4 (1979). Publications in Botany, No. 7. Ottawa:
National Museums of Canada, National Museum of Natural Sciences.

**Smith, Huron H.**
(1928). *Ethnobotany of the
Meskwaki Indians.* Bulletin of the
Public Museum of the City of
Milwaukee, Vol. 4, No. 2,
pp. 175–326.
(1932). *Ethnobotany of the Ojibwe
Indians.* Bulletin of the Public
Museum of the City of Milwaukee,
Vol. 4, No. 3, pp. 327–525.
(1933). *Ethnobotany of the Forest
Potawatomi Indians.* Bulletin of the
Public Museum of the City of
Milwaukee, Vol. 7, No. 1,
pp. 1–230.
(1970). *Ethnobotany of the Meno-
mini Indians.* Westport, Conn.:
Greenwood Press. Repr. of 1923
Bulletin of the Public Museum of
the City of Milwaukee, Vol. 4,
No. 1, pp. 1–174.

**Soper, J.H., and Margaret
Heimburger**
(1961). *100 Shrubs of Ontario.*
Toronto: Ontario Department of
Commerce and Development.

**Speck, Frank G., and Ralph W.
Dexter**
(1951). "Utilization of Animals and
Plants by the Micmac Indians of
New Brunswick". *Journal of the
Washington Academy of Sciences*,
Vol. 41, No. 8, pp. 250–59.

**Standley, P.C.**
(1943). *Edible Plants of the Arctic
Region.* Washington, D.C.: United
States Government Printing Office.

**Steedman, E.V., ed.**
(1930). "The Ethnobotany of the
Thompson Indians of British
Columbia". Pages 441–522 in
*Bureau of American Ethnology,
45th Annual Report, 1927–28.*
Washington, D.C.: Smithsonian
Institution.

**Stewart, Anne Marie, and Leon
Kronoff**
(1975). *Eating from the Wild.* New
York: Ballantine.

**Sturtevant, Edward L.**
(1972). *Sturtevant's Edible Plants
of the World.* Ed. by U.P. Hedrick.
New York: Dover. Repr. of 1919
ed., entitled *Sturtevant's Notes on
Edible Plants.* Albany, N.Y.: New
York Department of Agriculture
and Markets 27th Annual Report,
Vol. 2, Pt. 2.

**Szczawinski, Adam F., and George A. Hardy**
(1971). *Guide to Common Edible Plants of British Columbia*. British Columbia Provincial Museum Handbook No. 20. Victoria.

**Szczawinski, Adam F., and Nancy J. Turner**
(1978). *Edible Garden Weeds of Canada*. Ottawa: National Museums of Canada, National Museum of Natural Sciences.

**Tomikel, John**
(1973). *Edible Wild Plants of Pennsylvania and New York*. Pittsburgh: Allegheny Press.

**Turner, Nancy J.**
(1975). *Food Plants of British Columbia Indians*. Pt. 1, *Coastal Peoples*. British Columbia Provincial Museum Handbook No. 34. Victoria.
(1978). *Food Plants of British Columbia Indians*. Pt. 2, *Interior Peoples*. British Columbia Provincial Museum Handbook No. 36. Victoria.
(in press). *Plants in British Columbia Indian Technology*. British Columbia Provincial Museum Handbook. Victoria.

**Turner, Nancy J., and Adam F. Szczawinski**
(1978). *Wild Coffee and Tea Substitutes of Canada*. Ottawa: National Museums of Canada, National Museum of Natural Sciences.

**Underhill, J.E. (Ted)**
(1974). *Wild Berries of the Pacific Northwest*. Saanichton, B.C.: Hancock House.

**United States Department of Agriculture**
(1973). *Complete Guide to Home Canning, Preserving, and Freezing*. New York: Dover.

**Waugh, F.W.**
(1916). *Iroquois Foods and Food Preparation.* Canada Department of Mines, Geological Survey Memoir 86, Anthropological Series No. 12. Ottawa: Government Printing Bureau.

**Weiner, Michael A.**
(1972). *Earth Medicine—Earth Foods*. New York: Collier-Macmillan.

**Wittrock, Marion A., and G.L. Wittrock**
(1942). "Food Plants of the Indians". *Journal of the New York Botanical Garden*, Vol. 43, No. 507, pp. 57–71.

**Young, Steven B., and Edwin S. Hall**
(1965). "Contributions to the Ethnobotany of the St. Lawrence Island Eskimo". *Anthropological Papers of the University of Alaska*, Vol. 14, No. 2, pp. 43–53.

# Index